SpringerBriefs in Molecular Science

Protein Folding and Structure

Series editor

Cláudio M. Gomes, Oeiras, Portugal

About the Series

Prepared by leading experts, the series contains diverse types of contributions, from snapshot volumes that allow fast entry to a general topic to those covering more specialized aspects in the field of protein folding and structure. In common, these *Briefs* aim at covering essential concepts, methodologies, and ideas in the context of contemporary research in protein science. Through these compact volumes, this series serves as a venue for publication between typical research papers, review articles and full books, and aims at a broad audience, from students to researchers in academia and industry.

About the Editor

Cláudio M. Gomes joined the Gulbenkian Ph.D. program in Biology and Medicine (1994) and obtained his Ph.D. in Biochemistry (1999) from the Instituto de Tecnologia Química e Biológica, Universidade Nova de Lisboa. His research is focused on protein folding, misfolding and aggregation, mostly concerning molecular mechanisms in neurodegenerative and metabolic diseases. He is involved in extensive publishing and editorial activities, as an author in scientific journals, member of the Editorial boards of several scientific journals and editor of thematic journals issues and books.

More information about this series at http://www.springer.com/series/11958

Vladimir N. Uversky

Intrinsically Disordered Proteins

 Springer

Vladimir N. Uversky
Department of Molecular Medicine
University of South Florida
Tampa, FL
USA

ISSN 2199-3157   ISSN 2199-3165   (electronic)
ISBN 978-3-319-08920-1   ISBN 978-3-319-08921-8   (eBook)
DOI 10.1007/978-3-319-08921-8

Library of Congress Control Number: 2014945349

Springer Cham Heidelberg New York Dordrecht London

Printed on acid-free paper

Springer is part of Springer Science+Business Media (www.springer.com)

Foreword

The perspective of changing established paradigms is a fascinating aspect of the scientific endeavor. In structural biology, a long-standing paradigm has been the one that protein structure determines function and its reciprocal, that protein function is determined by structure. This view roots on the pioneering reports of protein structures determined using X-ray crystallography already over half a century ago, and has gained momentum with a wealth of protein three-dimensional structures which have been experimentally determined ever since. Undoubtedly, a well-folded protein structure is the hallmark attribute of protein function for many types of proteins, notably those adopting a globular fold. Nevertheless, today we also know that many proteins contain functional disordered segments or are wholly disordered, lacking well-defined three-dimensional structures, but are yet able to adopt an ensemble of functional conformations. Such proteins that challenge the structure-function paradigm are coined as intrinsically disordered, and are nowadays broadly accepted as important players in cellular functions. In this volume, the leading expert Vladimir N. Uversky provides us with an excellent overview on different aspects of this important area in protein structural biology. Through a direct, yet comprehensive presentation of the fundamental concepts, characteristics, and functions of intrinsically disordered proteins, along with valuable personal notes and historical insights, Uversky presents us with state-of-the-art knowledge and delivers a view of promising roles of intrinsically disordered proteins in biomedicine. As the Editor, I can hardly imagine a better inaugural volume for the *SpringerBriefs* series on *Protein Folding and Structure* than a volume dealing with *Intrinsically Disordered Proteins*, a paradoxical and debatable concept in the past of structural biology, but certainly one of the most promising future areas of protein research.

Oeiras, May 2014   Cláudio M. Gomes

Acknowledgements

Over the years, I collaborated with more than 700 colleagues from more than 100 universities in more than 20 countries. This work would be impossible without these numerous collaborators whose enthusiasm and help drove the studies on intrinsically disordered proteins for many years. In no particular order the list of people contributed to this project at its different stages includes: Eugene Permyakov, Joel Gillespie, Vyacheslav Abramov, Anthony Fink, Larissa Munishkina, Pierre Souillac, Sebastian Doniach, Ian Millett, Daniel Denning, Kevin Lee, Alexander Sigalov, Michael Rexach, Sergei Permyakov, Keith Oberg, Stefan Winter, Jie Li, Oxana Galzitskaya, Leonhard Kittler, Gunter Lober, Olga Tcherkasskaya, Seung-Jae Lee, Min Zhu, Amy Manning-Bog, Alison McCormack, Donato Di Monte, Kiowa Bower, I-Hsuan Liu, Gregory Cole, John Goers, Ghiam Yamin, Atta Ahmad, Charles Glaser, Elisa Cooper, Jeffrey Cohlberg, Mark Hokenson, Sudha Rajamani, Joseph Zbilut, Alessandro Giuliani, Alfredo Colosimo, Julie Mitchell, Mauro Colafranceschi, Norbert Marwan, Charles Webber, Jr., Christopher Oldfield, Yugong Cheng, Marc Cortese, Celeste Brown, Mikhail Karymov, Yuri Lyubchenko, Pedro Romero, Lilia Iakoucheva, Zoran Obradovic, Véronique Receveur-Bréchot, Jean-Marie Bourhis, Bruno Canard, Sonia Longhi, Chad McAllister, Yoshiko Kawano, Alexander Lushnikov, Andrew Mikheikin, Andrey Vartapetyan, Predrag Radivojac, Slobodan Vucetic, Timothy R. O'Connor, Jag Bhalla, Geoffrey Storchan, Caitlin MacCarthy, Maureen Harrington, Tanguy LeGall, Patrizia Polverino de Laureto, Laura Tosatto, Erica Frare, Oriano Marin, Angelo Fontana, Megan Sickmeier, Justin Hamilton, Vladimir Vacic, Agnes Tantos, Beata Szabo, Peter Tompa, Jake Chen, King Pan Ng, Gary Potikyan, Rupert Savene, Christopher Denny, Vinay Singh, Yue Zhou, Joseph Marsh, Julie Forman-Kay, Jingwen Liu, Zongchao Jia, Millie Georgiadis, Amrita Mohan, Ann Roman, Jiangang Liu, Narayanan Perumal, Eric Su, Fei Ji, Niels Klitgord, Michael Cusick, Marc Vidal, Chad Haynes, Saima Zaidi, Ya Yin Fang, Jessica Chen, Igor Lednev, Bin Xue, Leo Breydo, Madan Babu, Richard Kriwacki, Ming Xu, Vladimir Ermolenkov, Vitali Sikirzhytski, Natalya Topilina, Gaius Takor, Seiichiro Higashiya, John Welch, Konstantin Turoverov, Irina Kuznetsova, Dickran Aivazian, Lawrence Stern, Hongbo Xie, Liwei Li, Samy Meroueh, Garret Anderson, Rafael Lujan,

Arthur Semenov, Marco Pravetoni, Joseph Song, Ching-Kang Chen, Kevin Wickman, Kirill Martemyanov, Ekaterina Posokhova, Alexey Uversky, William Sullivan, Ágnes Tóth-Petróczy, Bálint Mészáros, István Simon, Monika Fuxreiter, Jeffrey Hansen, Francisco Asturias, Yuichiro Takagi, Oleg Paliy, Shawn Gargac, Alfredo De Biasio, Corrado Guarnaccia, Matija Popovic, Alessandro Pintar, Sándor Pongor, Mona Rahman, Kim Munro, Steven Smith, Ivan Pacheco, John MacLeod, Anush Bakunts, Alexander Denesyuk, Ekaterina Knyazeva, Ramil Ismailov, Gerard Goh, Kimmo Rantalainen, Perttu Permi, Nisse Kalkkinen, Kristiina Mäkinen, Siyuan Ren, Zhengjun Chen, Fabien Durand, Adilia Dagkessamanskaia, Helene Martin-Yken, Marc Graille, Herman Van Tilbeurgh, Matteo Binda, Frédéric Lopez, Karim El Azzouzi, Jean-Marie Francois, Andrew Campen, Ryan Williams, Joel Sussman, Israel Silman, Dong-Pyo Hong, Nobuhiko Tokuriki, Igor Berezovsky, Dan Tawfik, Eugénie Hébrard, Yannick Bessin, Thierry Michon, François Delalande, Alain Van Dorsselaer, Jocelyne Walter, Nathalie Declerk, Denis Fargette, Wenbo Zhou, Carlo Santambrogio, Stefania Brocca, Rita Grandori, Mária Šamalíková, Marina Lotti, Marco Vanoni, Lilia Alberghina, Lorenzo Testa, Annalisa D'Urzo, Johnny Habchi, Uros Midic, Bo He, Kejun Wang, Yunlong Liu, Xiaoyun Meng, David Eliezer, Victor Kutyshenko, Dmitry Prokhorov, Maria Timchenko, Yuri Kudrevatykh, Liubov' Gushchina, Vladimir Khristoforov, Vladimir Filimonov, Agyakojo Frimpong, Rinat Abzalimov, Igor Kaltashov, Roland Dunbrack, Robert Williams, Xiaolin Sun, William Jones, Erik Rikkerink, David Greenwood, Matthew Templeton, David Libich, Tony McGhie, Minsoo Yoon, Wei Cui, Christopher Kirk, Dawn Harvey, Patrick Edwards, Steven Pascal, Christopher Kirk, Thérèse Considine, David Sheerin, Jasna Rakonjac, Wei-Lun Hsu, Jun-Ho Lee, Hua Lu, Rajarshi Ghosh, Tatiana Nikitina, Rachel Horowitz-Scherer, Lila Gierasch, Kristopher Hite, Jeffrey Hansen, Christopher Woodcock, Prerna Malaney, Ravi Pathak, Vrushank Davé, Jaymin Kathiriya, Eric Clayman, Aaron Satner, Julian Jorda, Andrey Kajava, Hayriye Erkizan, Jeffrey Toretsky, Justin Yamada, Joshua Phillips, Samir Patel, Gabriel Goldfien, Alison Calestagne-Morelli, Hans Huang, Ryan Reza, Justin Acheson, Viswanathan Krishnan, Shawn Newsam, Ajay Gopinathan, Edmond Lau, Michael Colvin, Olof Einarsdottir, Blanca Silva, Stacy Dixon, Micah Bhatti, Natalia Moroz, Stefanie Novak, Ricardo Azevedo, Mert Colpan, Carol Gregorio, Alla Kostyukova, Samar Shah, Yulia Gritsyna, Sarah Hitchcock-DeGregori, Mark Rochman, Leila Taher, Toshihiro Kurahashi, Srujana Cherukuri, David Landsman, Ivan Ovcharenko, Michael Bustin, Marcin Mizianty, Tuo Zhang, Yaoqi Zhou, Zhenling Peng, Fatemeh Miri Disfani, Lukasz Kurgan, Xiao Fan, Patrick Dolan, Douglas LaCount, Daniel Soeria-Atmadja, Mats Gustafsson, Ulf Hammerling, Franck Peysselon, Sylvie Ricard-Blum, Barbara Zambelli, Nunilo Cremades, Paolo Neyroz, Paola Turano, Stefano Ciurli, Fernanda Lopes, Olena Dobrovolska, Rafael Guerra, Valquiria Broll, Célia Carlini, Fei Huang, Colin Burns, Tuo Zhang, Eshel Faraggi, Bogdan Melnik, Tatiana Povarnitsyna, Anatoly Glukhov, Tatyana Melnik, Sandy Westerheide, Rachel Raynes, Chase Powell, Kuiran Xu, Ya-Yue Van, Maria Noutsou, Madelon Maurice,

Stefan Rüdiger, Albert William, Srinivas Tipparaju, Xiao-Ping Li, Peter J. Kilfoil, Aruni Bhatnagar, Oleg Barski, Rubén Hervás, Javier Oroz, Albert Galera-Prat, Oscar Goñi, Alejandro Valbuena, Andrés Vera, Àngel Gómez-Sicilia, Fernando Losada-Urzáiz, Margarita Menéndez, Douglas Laurents, Marta Bruix, Mariano Carrión-Vázquez, Carrie Croy, Farha Vasanwala, Derrick Johnson, Xiaoyue Zhao, Aaron Friedman, Phineus Markwick, Andrew McCammon, Alexander Kirilyuk, Mika Shimoji, Jason Catania, Geetaram Sahu, Nagarajan Pattabiraman, Antonio Giordano, Christopher Albanese, Italo Mocchetti, Maria Laura Avantaggiati, Mark Howell, Ryan Green, Alexis Killeen, Lamar Wedderburn, Vincent Picascio, Alejandro Rabionet, Maya Larina, Matt Oates, Takashi Ishida, Mohamed Ghalwash, Zsuzsanna Dosztányi, Julian Gough, Eldon Ulrich, John Markley, Maria Ribeiro, Julio Espinosa, Sameen Islam, Osvaldo Martinez, Jayesh Thanki, Stephanie Mazariegos, Tam Nguyen, Juan Ortiz1, Madolyn MacDonald, Patrick Masterson, Jessica Siltberg-Liberles, Georgina Rae, Karine David, Marion Wood, Rong-Mei Wu, Eric F. Walton, Roger Hellens, Elisar Barbar, Martin Blackledge, Sarah Bondos, Jane Dyson, Jörg Gsponer, Kyou-Hoon Han, David Jones, Steven Metallo, Ken Nishikawa, Ruth Nussinov, Rohit Pappu, Burkhard Rost, Philipp Selenko, Vinod Subramaniam, Francois-Xavier Theillet, Lajos Kalmar, Gary Daughdrill, Ariel Azia, Amnon Horovitz, Ron Unger, Umesh Jinwal, Elias Akoury, Jose Abisambra, John O'Leary, Andrea D. Thompson, Laura Blair, Ying Jin, Justin Bacon, Bryce Nordhues, Matthew Cockman, Juan Zhang, Pengfei Li, Bo Zhang, Sergiy Borysov, Jacek Biernat, Eckhard Mandelkow, Jason Gestwicki, Markus Zweckstetter, Chad Dickey, Shannon Hill, Matthew Scaglione, Sarah Fontaine, Li Wang, Carl Cotman, Henry Paulson, Martin Muschol, Torsten Klengel, Elisabeth Binder, Rakez Kayed, Todd Golde, Nicole Berchtold, Jing Yan, Paul Filipow, Poornima Bhat-Nakshatri, Eun-Kyung Song, Nikail Collins, Bert O'Malley, Tim Geistlinger, Jason Carroll, Myles Brown, Harikrishna Nakshatri, Ucheor Choi, Rashek Kazi, Natalie Stenzoski, Lonnie Wollmuth, Mark Bowen, Insung Na, Krishna Reddy, Derek Redmon, Markus Kopa, Yiru Qin, Larisa Mikheeva, Pedro Madeira, Boris Zaslavsky, Macarena Marín, Thomas Ott, Shelly DeForte, Chewook Lee, Alessandro Piai, Isabella Felli, Roberta Pierattelli, Mihaly Varadi, Simone Kosol, Pierre Lebrun, Erica Valentini, Dmitri Svergun, Michele Vendruscolo, David Wishart, Peter Wright, Ketaki Ganti, Lawrence Banks, Yoshihisa Sakai, Bernd Sokolowski, Carrie Butler, Olivier Lucas, Stefan Wuchty, Hana Popelka, Daniel Klionsky, Gabor Kovacs, Leonid Breydo, Ryan Green, Viktor Kis, Gina Puska, Péter Lőrincz, Laura Perju-Dumbrava, Regina Giera, Walter Pirker, Mirjam Lutz, Ingolf Lachmann, Herbert Budka, Kinga Molnár, Lajos László, Elrashdy Redwan, Esmail EL-Fakharany, Mustafa Linjawi, Ursula Jacob, Robin van der Lee, Marija Buljan, Benjamin Lang, Robert J. Weatheritt, Philip Kim, Andreas Joerger, Daniel Kraut, Andreas Matouschek, Roderick Lim, David Blocquel, Vikas Pejaver, Fuxiao Xin, and Cláudio Gomes. This is my interactome and I am grateful to all former and current colleagues for their priceless contributions, assistance, and support. I would like also to thank my family: my mother Galina, sisters Marianna and Tamara, and sons Alexey and Sergei, for their unending love and support. Last, but not least, my love and deepest gratitude go to my loving wife, Elena, for her understanding, dedication, support, endless patience, inspiration, and constant encouragement, without which this work would not be possible.

Vladimir N. Uversky

Contents

About the Author

**Vladimir N. Uversky** obtained his Ph.D. in Biophysics from Moscow Institute of Physics and Technology (1991) and D.Sc. in Biophysics from Institute of Experimental and Theoretical Biophysics, Russian Academy of Sciences (1998). He spent his early career working on protein folding at Institute of Protein Research and the Institute for Biological Instrumentation (Russian Academy of Sciences). In 1998, he moved to the University of California, Santa Cruz to work on protein folding, misfolding, and protein intrinsic disorder. In 2004, he moved to the Center for Computational Biology and Bioinformatics at the Indiana University Purdue University Indianapolis to work on the intrinsically disordered proteins. Since 2010, he is with the Department of Molecular Biology at the University of South Florida.

Abstract

Nothing is solid about proteins. Governing rules and established laws are constantly broken. As an example, the last decade and a half has witnessed the fall of one of the major paradigms in structural biology. Contrary to the more than century-old belief that the unique function of a protein is determined by its unique structure, which, in its turn, is defined by the unique amino acid sequence, many biologically active proteins lack stable tertiary and/or secondary structure either entirely or at their significant parts. Such Intrinsically Disordered Proteins (IDPs) and hybrid proteins containing ordered domains and functional IDP regions (IDPRs) are highly abundant in nature, and many of them are associated with various human diseases. Such disordered proteins and regions are very different from ordered and well-structured proteins and domains at a variety of levels and possess well-recognizable biases in their amino acid compositions and amino acid sequences. A characteristic feature of these proteins is their exceptional structural heterogeneity, where different parts of a given polypeptide chain can be ordered (or disordered) to different degrees. As a result, a typical IDP/IDPR contains a multitude of potentially foldable, partially foldable, differently foldable, or not foldable structural segments. This distribution of conformers is constantly changing in time, where a given segment of a protein molecule has different structures at different time points. The distribution is also constantly changing in response to changes in the environment. This mosaic structural organization is crucial for their functions and many IDPs are engaged in biological functions that rely on high conformational flexibility and that are not accessible to proteins with unique and fixed structures. As a result, the functional repertoire of IDPs complements that of ordered proteins, with IDPs/IDPRs being often involved in regulation, signaling, and control. This SpringerBriefs volume is dedicated to IDPs and IDPRs and an attempt is made to compress a massive amount of knowledge and into a digest that aims to be of use to those wishing a fast entry into this promising field of structural biology.

Intrinsically Disordered Proteins

1 Introducing the Phenomenon of Protein Intrinsic Disorder

A typical practical approach to deal with non-orthodox scientific findings is to neglect and ignore them. In essence, this is an ostrich-head-in-the-sand way of avoiding unknowns. As a result of this approach, for more than a century, the protein-related research was mostly focused on proteins with unique structures. Although some degree of flexibility was allowed here and there in a protein molecule, the concept one sequence—one structure—one function ruled thoughts of the researchers since 1894, when Emil Fischer proposed his famous "lock-and-key" model to describe the molecular mechanism of enzymatic activity [1]. This viewpoint was further solidified by the successful solution of X-ray crystallographic structures of many proteins. In fact, as of January 7, 2014 there were 94,153 structures of proteins and protein-nucleic acid complexes in the Protein Data Bank [2], with 84,084 of these structures (89.3 %) being determined by X-ray crystallography. It was pointed out that these many crystal structures reinforced a static view of functional protein, where a rigid active site of an enzyme can be seeing as a sturdy lock that provides an exact fit to only one key, a specific and unique substrate [3–5]. It was also pointed out that despite numerous limitations, this "lock-and-key" model was an extremely fruitful concept that was responsible for the creation of the modern protein science [3, 5]. This fact is illustrated by Fig. 1a that shows some of the most obvious scientific consequences of the application of structure-function paradigm which is deservedly placed at the center of the 'Big Bang' model that gives rise to the universe of protein science [3, 5].

Obviously, because of the cornerstone "lock-and-key" paradigm, this original protein science universe dealt mostly with ordered proteins characterized by a well-defined structure. However, even the most structured proteins are not rigid crystal-like entities but should be considered as dynamic systems possessing different degree of conformational flexibility [5]. In fact, conformational forces stabilizing

© The Author(s) 2014

V.N. Uversky, *Intrinsically Disordered Proteins*, Protein Folding and Structure, DOI 10.1007/978-3-319-08921-8_1

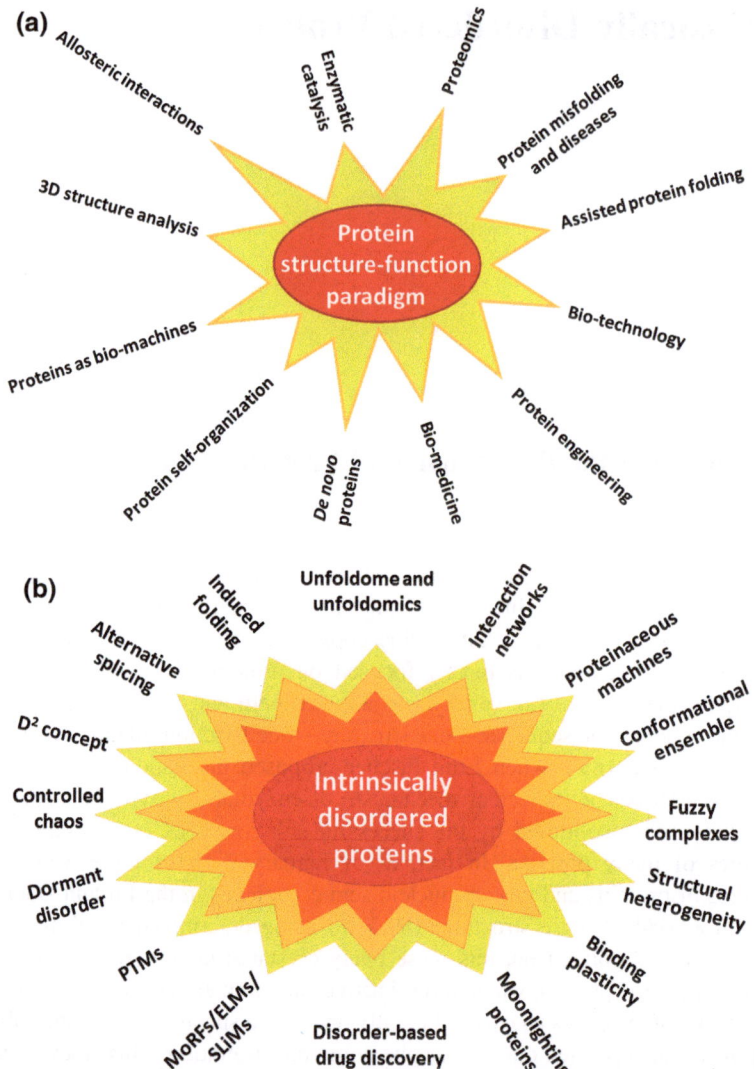

Fig. 1 **a** Protein structure-function paradigm can be considered as the 'Big Bang' created universe of the modern protein science. Some major directions based on the consideration of protein function as lock-and-key mechanism are shown. Modified from Ref. [3]. **b** Paradigm shift caused by the introduction of the protein intrinsic disorder concept opened a wide array of new directions in protein science. In essence, introduction of this concept can be considered as a scientific revolution that, according to Kuhn [314], "occurs when scientists encounter anomalies that cannot be explained by the universally accepted paradigm within which scientific progress has thereto been made" (http://en.wikipedia.org/wiki/Paradigm_shift). This figure is reproduced from Ref. [5]

the protein structure are weak and can be broken even at ambient temperatures due to the thermal fluctuations [4, 5]. This provides groups involved in such interactions with the ability to be form new weak interactions of comparable energy [4]. In ordered proteins, such structural rearrangements are of relatively small scale and are relatively fast. As a result, the 3-D structures of proteins determined by X-ray crystallography and many other physical techniques represent averaged pictures [6].

It was pointed out [4] that in addition to ordered proteins possessing disordered regions of varying length, literature contains numerous examples of biologically active proteins with flexible structures. Therefore, there is another class of functional proteins and protein regions that contain shorter of longer regions with high conformational dynamics, and some proteins are even characterized by a complete or almost complete lack of ordered structure under physiological conditions (at least in vitro) which appears to be a critical aspect of these proteins' function in vivo [4, 7–13]. These proteins and protein regions (which are known now as intrinsically disordered proteins (IDPs) and intrinsically disordered protein regions (IDPRs)) have no single, well-defined equilibrium structure and exist as heterogeneous ensembles of conformers such that no single set of coordinates or backbone Ramachandran angles is sufficient to describe their conformational properties. IDPs and hybrid proteins, that in addition to ordered domains contains various IDPRs [13], display marked conformational heterogeneity and constitute a significant part of the protein kingdom [14–17].

These proteins were independently discovered one-by-one over a long period of time and therefore they were considered as rare exceptions to the general rule [5]. Although the phenomenon of biological functionality without stable structure was rediscovered multiple times, for a long time it was unnoticed by a wide audience because the authors frequently invented new terms to describe their protein of interest [13]. In fact, an incomplete list of terms coined in the literature to describe these proteins includes floppy, pliable, rheomorphic [18], flexible [19], mobile [20], partially folded [21], natively denatured [22], natively unfolded [9, 23], natively disordered [12], intrinsically unstructured [8, 11], intrinsically denatured [22], intrinsically unfolded [23], intrinsically disordered [10], vulnerable [24], chameleon [25], malleable [26], 4D [27], protein clouds [28], dancing proteins [29], proteins waiting for partners [30], and several other names often representing different combinations of 'natively/naturally/inherently/intrinsically' with 'unfolded/unstructured/disordered/denatured' among several others. Therefore, majority of the names used in the early literature express that the 'unfolded, unstructured, disordered, and denatured' state is a 'native, natural, inherent, and intrinsic' property of these proteins [13].

Although protein intrinsic disorder is considered now as an established concept and PubMed contains hundreds and hundreds of papers dealing with different aspects of IDPs/IDPRs, the route to recognition of these proteins as a novel functional entity was complex and lengthy. As it is often the case for new scientific concepts, the idea of structure-less functionality went through the stages of passive ignorance (ostrich-head-in-the-sand) and active denial (this is impossible because this never could happen) to scrupulous examination and enthusiastic acceptance. For example, it took me more than a year to publish my first paper dedicated to the systematic analysis of such proteins, and the manuscript was successively rejected by 14 journals before it was finally accepted by Proteins [9]. However, it is clear now that the concept of protein intrinsic disorder is a useful invention and could be considered as a universal lock-pick that helps solving many of the seemingly unsolvable problems in protein science [5]. One could say that this idea gave a new boost to the development of the protein science, generating a wide array of principally novel research directions (see Fig. 1b) [5].

2 Peculiar Amino Acid Sequences of Soluble IDPs

Identification of IDPs as unique entities belonging to a new protein tribe is directly related to the recognition that the amino acid sequences of these proteins are dramatically different from sequences of ordered proteins [7, 9, 10, 31–33]. In fact, at given conditions, the capability of a particular polypeptide chain to fold into a unique biologically active structure or to stay as biologically active conformational ensemble is encoded in the specific features of its amino acid sequence. For example, it has been pointed out that the low content of hydrophobic residues combined with the high content of similarly charged residues represents a characteristic feature of some IDPs (so called extended IDPs or natively unfolded proteins with coil-like or close to coil-like structures) [9]. Therefore, compact proteins and extended IDPs can be distinguished based only on their net charges and hydropathies using a simple charge-hydropathy (CH) plot, where the natively unfolded proteins are specifically localized within a specific region of CH phase space and are reliably separated from compact ordered proteins [9].

More detailed comparison of amino acid sequences revealed that in comparison with ordered proteins and domains, the IDPs/IDPRs are significantly depleted in so-called order-promoting amino acids which are bulky aliphatic (Ile, Leu, and Val) and aromatic amino acid residues (Trp, Tyr, and Phe) that would normally form the hydrophobic core of a folded globular protein, and also possess low content of Cys and Asn residues [7, 34]. Instead, IDPs/IDPRs are enriched in disorder-promoting residues such as Ala, polar amino acids Arg, Gly, Gln, Ser, Glu, and Lys, and also contain high proportions of the hydrophobic, but structure-breaking Pro [10, 32, 33, 35, 36].

Difference between ordered and disordered proteins goes far beyond described above differences in their amino acid compositions. In fact, based on the comparison of the 265 amino acid physico-chemical property-based scales (such as hydropathy, net charge, flexibility index, helix propensities, strand propensities, aromaticity, etc.) [35] and more than 6,000 composition-based attributes (e.g., all possible combinations having one to four amino acids in the group) [37] it has been concluded that ordered and disordered proteins and regions can be discriminated using at least ten of these attributes, such as 14 Å contact number, hydropathy, flexibility, β-sheet propensity, coordination number, content of major disorder-promoting residues (Arg + Ser + Pro + Glu), bulkiness, content of major order-promoting residues (Cys + Trp + Tyr + Phe), volume, and net charge [10]. Later, a novel amino acid scale to discriminate between order and disorder was built based on the analysis of 517 amino acid scales (such as a variety of hydrophobicity scales, different measures of side chain bulkiness, polarity, volume, compositional attributes, the frequency of each single amino acid and so on) [31]. This scale provided a new ranking for the tendencies of the amino acid residue to promote order or disorder (from order promoting to disorder promoting): Trp, Phe, Tyr, Ile Met, Leu, Val, Asn, Cys, Thr, Ala, Gly, Arg, Asp, His, Gln, Lys, Ser, Glu, and Pro [31].

The fact that the sequences of ordered and disordered proteins and regions are noticeably different suggested that IDPs clearly constitute a separate entity inside the protein kingdom and that these proteins can be reliably predicted using various computational tools [38–43]. Furthermore, IDPs/IDPRs and ordered globular proteins/domains are expected be very different structurally since peculiarities of amino acid sequence determine peculiarities of protein structure [5].

3 Natural Abundance of IDPs/IDPRs

Already first systematic analyses revealed that intrinsic disorder in proteins is a rather common phenomenon. In fact, as of 2002, the list of experimentally validated natively unfolded proteins with chain length greater than 50 amino acid residues contained more than 100 entries [3, 44]. It was also pointed out that this list would probably be doubled if shorter polypeptides 30–50 residues long were included [3], and that these 100 experimentally validated natively unfolded have at least 250 homologues, which are also expected to be natively unfolded [3, 9]. It happened that these 'large' numbers (which actually were large enough to make a crucial point that biologically active structure-less proteins are new rule and not mere rare exceptions) constitute just a small tip of an iceberg. In fact, using computational tools developed for the sequence-based intrinsic disorder prediction (such family of predictors of naturally disordered regions, PONDRs [7, 10, 14, 45], and many other computational tools [38–42]) the wide spread of IDPs and hybrid proteins with IDPRs was convincingly shown. For example, more than 15,000 out of 91,000 proteins in the then-current Swiss Protein database were identified as having long IDRs [45], and the published in 2000 analysis of 31 whole genomes that span the 3 kingdoms of life revealed that many proteins contained segments predicted to have ≥40 consecutive disordered residues [14]. In that study, eukaryotes were shown to exhibit more disorder by this measures than either prokaryotes or archaea, with C. elegans, A. thaliana, S. cerevisiae, and D. melanogaster predicted to have 52–67 % of their proteins with such long IDPRs, whereas bacteria and archaea were predicted to have 16–45 and 26–51 % of their proteins with long IDPRs, respectively [14]. Other studies on the fraction of intrinsic disorder in various evolutionary distant species supported these findings and consistently showed that the eukaryotic proteomes had a higher fraction of intrinsic disorder than prokaryotic proteomes [15, 46–50]. These observations were explained based on the specific functional repertoire of IDPs/IDPRs, which are often involved in signaling, recognition, and regulation, and there are complex and well-developed regulation networks in eukaryotic and especially in muticellular eukaryotic organisms that might rely on the ability of IDPs/IDPRs to perform the necessary functions [4, 16, 51, 52].

The increased abundance of disorder in higher eukaryotes was repeatedly shown by studies which progressively analyzed more and more completed proteomes, finally culminating with a comprehensive investigation of the disorder distribution in almost 3,500 proteomes from viruses and three kingdoms of life [17]. Figure 2 represents the results of this analysis by showing the correlation between the intrinsic disorder content and proteome size for 3,484 species from viruses, archaea, bacteria, and eukaryotes.

Surprisingly, Fig. 2 shows that there is a well-defined gap between prokaryotes and eukaryotes in the plot of fraction of disordered residues on proteome size, where almost all eukaryotes have 32 % or more disordered residues, whereas the majority of the prokaryotic species have 27 % or fewer disordered residues [17]. Therefore, it looks like the fraction of 30 % disordered residues serves as a

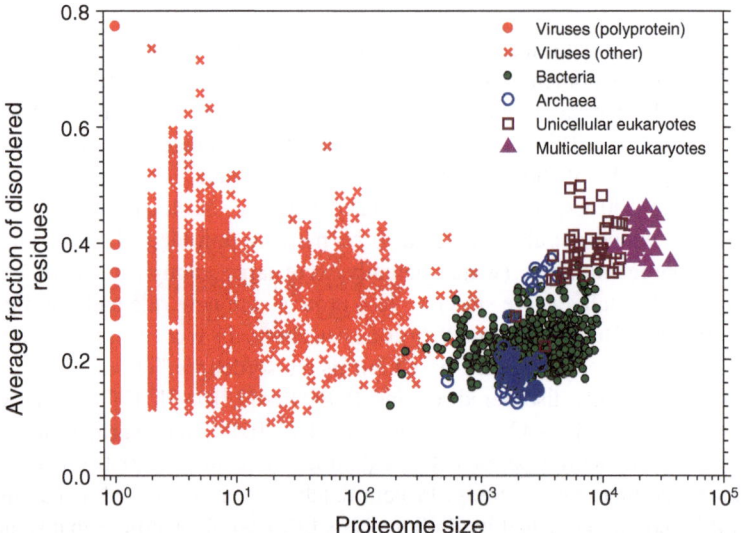

Fig. 2 Distribution of intrinsic disorder in various proteomes. Correlation between the intrinsic disorder content and proteome size for 3,484 species from viruses, archaea, bacteria, and eukaryotes. Each symbol indicates a species. There are totally six groups of species: viruses expressing one polyprotein precursor (*small red circles filled with blue*), other viruses (*small red circles*), bacteria (*small green circles*), archaea (*blue circles*), unicellular eukaryotes (*brown squares*), and multicellar eukaryotes (*pink triangles*). Each viral polyprotein was analyzed as a single polypeptide chain, without parsing it into the individual proteins before predictions. The proteome size is the number of proteins in the proteome of that species and is shown in log base. The average fraction of disordered residues is calculated by averaging the fraction of disordered residues of each sequence over the all sequences of that species. Disorder prediction is evaluated by PONDR-VSL2B. This figure is reproduced from Ref. [5]

boundary between the prokaryotes and eukaryotes and reflects the existence of a complex step-wise correlation between the increase in the organism complexity and the increase in the amount of intrinsic disorder. A gap in the plot of fraction of disordered residues on proteome size parallels a morphological gap between pro-karyotic and eukaryotic cells which contain many complex innovations that seemingly arose all at once. In other words, this sharp jump in the disorder content in proteomes associated with the transition from prokaryotic to eukaryotic cells suggests that the increase in the morphological complexity of the cell paralleled the increased usage of intrinsic disorder [17]. The variability of disorder content in unicellular eukaryotes and rather weak correlation between disorder status and organism complexity (measured as the number of different cell types) is likely related to the wide variability of their habitats, with especially high levels of dis-order being found in parasitic host-changing protozoa, the environment of which changes dramatically during their life-span [53]. Further support for this hypothesis came from the fact that the intrinsic disorder content in multicellular eukaryotes (which are characterized by more stable and less variable environment of individual cells) was noticeably less variable than that in the unicellular eukaryotes [17].

4 Wavy Evolution of Intrinsic Disorder: Reinventing the Wheel

Results of computational studies on the abundance of intrinsic disorder at proteome level systematically show that IDPs/IDPRs are more common in eukaryotes than in less complex organisms [14, 15, 46–50]. Such compelling findings suggest that disorder, with its ability to be implemented in various signaling, recognition, and regulation pathways and networks, is important for the maintenance of life in eukaryotic and especially multicellular eukaryotic organisms [4, 16, 51, 52]. Also, the finding that alternatively spliced regions of mRNA code for IDPRs much more often than for structured regions suggested that there is a linkage between alter-native splicing and signaling by IDPRs that constitutes a plausible mechanism that could underlie and support cell differentiation, which ultimately gave rise to the multicellular eukaryotic organisms [54].

Based on these and other observations, one can assume that intrinsic disorder represents a relatively recent evolutionary invention. However, this hypothesis obviously would be wrong if earlier stages of evolution would be taken into account. In fact, the chances that the first polypeptides that appeared in the pri-mordial soup of the primitive Earth possessed well-developed and unique 3D structures are minimal. The Earth formed about 4.5 billion years ago. Scientists dated the first fossils to 3.85 billion years ago. There are still debates and different theories about what happened in those years between the time the earth was cool enough to spawn life and the time the first fossils were formed. At the beginning of the 20th century, Oparin [55] and Haldane [56] proposed that some organic molecules could have been spontaneously produced from the gases of the primitive Earth atmosphere assuming that this primitive atmosphere was reducing (as opposed to oxygen-rich), and there was an appropriate supply of energy, such as lightning or ultraviolet light. Thirty year later, this hypothesis (that constitute a cornerstone of the theory of molecular evolution) received strong support from the elegant experiments of Stanley L. Miller and Harold C. Urey who were able to synthesize various organic compounds including some amino acids from non-organic compounds which were believed to represent the major components of the early Earth's atmosphere (water vapor, hydrogen, methane, and ammonia) by putting them into a closed system and running a continuous electric current through the system, to simulate lightning storms believed to be common on the early Earth [57, 58]. However, the Miller-Urey experiment yielded only about half of the modern amino acids [57, 58], suggesting that the first proteins on Earth may have contained only a few amino acids.

These findings go in parallel with the biosynthetic theory of the genetic code evolution suggesting that the genetic code evolved from a simpler form that encoded fewer amino acids [59], probably paralleled by the invention of biosynthetic pathways for new and chemically more complex amino acids [60]. Furthermore, some additional support of the validity of this hypothesis can be found in the standard genetic code (that consists of $4 \times 4 \times 4 = 64$ triplets of nucleotides, codons), which is known to be redundant (64 codons encodes for 20 amino acids). In fact, with only two exceptions, codons encoding one amino acid may differ in any of their three positions. However, only the third positions of some codons may be fourfold degenerate; i.e., any nucleotide at this position specifies the same amino acid and all nucleotide substitutions at this site are synonymous. Using these observations as a reflection of the evolutionary development, it was proposed that there was a period during code evolution where the third position was not needed at all and a doublet code preceded the triplet code, giving rise to $4 \times 4 = 16$ codons (encoding for 16 or fewer amino acids, if a termination codon is taken into account) [61].

Therefore, based on these and many other premises, one can discriminate evolutionary old and new amino acids. In 2000, Eduard N. Trifonov combined 40 different of these single-factor criteria into a consensus and proposed the following temporal order of addition for the amino acids: Gly/Ala, Val/Asp, Pro, Ser, Glu/Leu, Thr, Arg, Asn, Lys, Gln, Ile, Cys, His, Phe, Met, Tyr, Trp [62]. Even superficial analysis of this sequence reveals that many of the early amino acids (such as Gly, Asp, Glu, Pro, and Ser) are disorder-promoting, as they are very abundant in modern intrinsically disordered proteins. On the other hand, the major order-promoting residues (Cys, Trp, Tyr, and Phe) were added to the genetic code late. This observation is further illustrated by Fig. 3a which represents modern genetic code and which contains information on the early and late codons (shown by light red and light blue colors respectively) and corresponding disorder- and order-promoting residues (shown by red and blue colors, respectively). Figure 3a shows that there is relatively good agreement between the "age" of the residue and its disorder-promoting capacity, with early residues being mostly disorder-

(a)

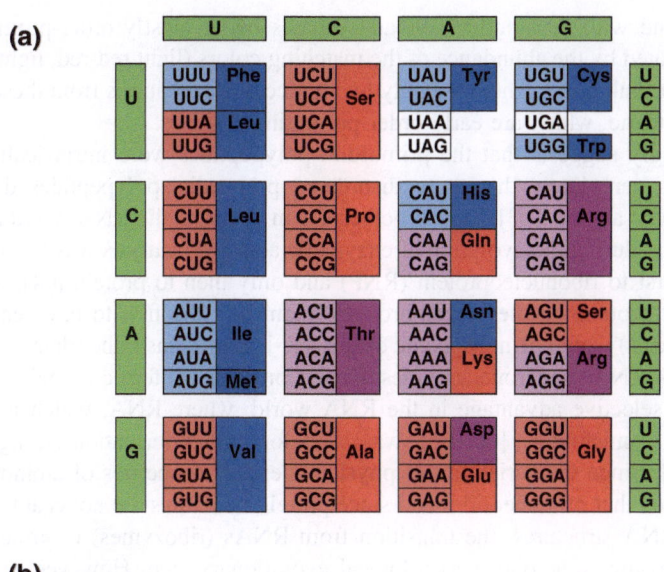

(b)

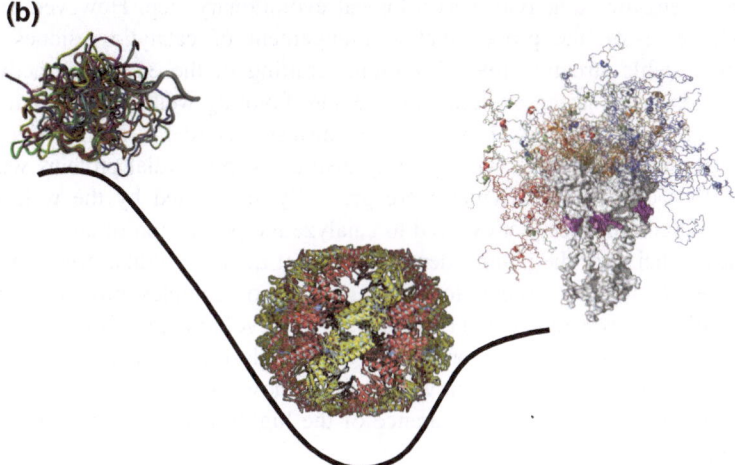

Fig. 3 Peculiarities of protein intrinsic disorder evolution. **a** Modern genetic code with information on the early and late codons (shown by *light red* and *light blue colors*, respectively) and disorder- and order-promoting residues (shown by *red* and *blue colors*, respectively). Intermediate codons are shown by *light pink color*. Disorder-neutral residues are shown by *pink color*. **b** *Wavy pattern* of the global disorder evolution. X-axis represents evolutionary time and Y-axis shows disorder content in proteins at given evolutionary time point. Here, primordial proteins are expected to be mostly disordered (*left-hand side* of the plot), proteins in LUA likely are mostly structured (*center* of the plot), whereas many protein in eukaryotes are either totally disordered or hybrids containing both ordered and disordered regions (*right-hand side* of the plot). This figure is reproduced from Ref. [5]

promoting, and with the majority of late residues being mostly order-promoting. This is illustrated by the abundance of the matching colors (light red-red, light blue-blue and light pink-pink). There are only two noticeable exceptions from these rule, valine and leucine, which are early order-promoting residues.

This strongly suggests that the primordial polypeptides were intrinsically disordered. It is very likely that these disordered primordial polypeptides did not possess catalytic activity [63]. This hypothesis is in line with the RNA world theory suggesting that during the evolution of enzymatic activity, catalysis was transferred from RNA first to ribonucleoprotein (RNP) and only then to protein [64]. Therefore, the first proteins in the "breakthrough organism" (the first to have encoded protein synthesis) would be nonspecific chaperone-like proteins rather than catalytic [63, 65]. Such RNA chaperone activities of early proteins conferred to their carriers a significant selective advantage in the RNA world, where RNA, which is especially prone to misfolding [66, 67], was used for both information storage and catalysis [68]. Since the variability of physico-chemical properties of amino acids greatly exceeds that of nucleotides and since protein structures are noticeably more stable than RNA structures, the transition from RNAs (ribozymes) to proteins as carriers of enzymatic activity was a logical evolutionary step. However, efficient catalysis relies on the proper spatial arrangement of catalytic residues which requires a stable structure [69]. Therefore, grafting of the enzymatic activity to proteins generated strong evolutionary pressure favoring well-folded structures.

In other words, the global evolution of intrinsic disorder is characterized by a wavy pattern (see Fig. 3b), where highly disordered primordial proteins with primarily RNA-chaperone activities were gradually substituted by the well-folded, highly ordered enzymes that evolved to catalyze the production of all the complex "goodies" crucial for the independent existence of the first cellular organisms. Due to its specific features crucial for the regulation of complex processes, protein intrinsic disorder was reinvented at the subsequent evolutionary steps leading to the development of more complex organisms from the last universal ancestor (i.e., the most recent organism from which all organisms now living on Earth descend [70, 71]), and culminating in the appearance of the highly elaborated eukaryotic cells (see Fig. 3b).

Diversified evolution of IDPs in modern organisms There is no simple answer to the question on the comparative evolutionary rates of ordered and intrinsically disordered proteins and regions in modern organisms. In fact, it looks like that everything is possible, and intrinsically disordered sequences may evolve faster, slower or similarly to ordered sequences. For example, disordered and ordered domains of the same protein (e.g., papillomavirus E7 oncoprotein) were shown to possess similar degrees of conservation and co-evolution [72]. Many other intrinsically disordered sequences were shown to be characterized by high evolutionary rates [73–75] determined by the lack of specific structural restrictions. In fact, the analysis of calcineurins [7], topoisomerase [76], ribosomal protein S4 [77], β-subunits of the potassium channel Kvβ1.1 [78], and many other proteins showed that disordered regions in these proteins contained more amino acid substitutions, insertions and deletions than the ordered regions of the same proteins [73, 79].

Furthermore, based on the observation that a significantly higher degree of positive Darwinian selection was observed in IDPRs of proteins compared to regions of α-helix, β-sheet or tertiary structures, it was hypothesized that IDPRs may be required for the genetic variation with adaptive potential and that these regions may be of "central significance for the evolvability of the organism or cell in which they occur" [80].

On the other hand, some IDPs and IDPRs are highly conserved. Human α-synuclein (a canonical IDP of 140 residues [81, 82]) differs from its mouse counterpart by mere six residues (4 %), and there are just 21 residue differences (12 %, which include residue differences at 18 positions and 3 insertions/deletions) between the human and canary α-synucleins [83]. In flagellin, the ordered central region has greater sequence diversity than its disordered termini [84]. Functionally important conserved regions of predicted disorder were shown to be rather common in proteins from all kingdoms of life, including viruses [85, 86]. Furthermore, many functional domains of a significant size were shown to be intrinsically disordered [87].

Overall, a systematic study of several families of proteins with at least one structurally characterized disordered region revealed that their IDPRs are characterized by highly heterogeneous evolutionary rates, with some disordered amino acid sequences evolving slowly, and others evolving more rapidly than ordered sequences [73]. Also, even different parts of the same disordered region can possess noticeable variability in their divergence during the evolutionary process [88]. Finally, in some disordered proteins, peculiarities of the amino acid composition, and not the amino acid sequence might be conserved [89, 90].

5 Structural Heterogeneity of IDPs: When Almost Everything Is Possible

One of the crucial consequences of an extended sequence space and non-homogeneous distribution of foldability (or the structure-coding potential) within amino acid sequences of IDPs and IDPRs is their astonishing structural heterogeneity. In fact, a typical IDP/IDPR contains a multitude of potentially foldable, partially foldable, differently foldable, or not foldable at all structural elements [91]. As a result, different parts of a molecule are ordered (or disordered) to a different degree, this distribution is constantly changing in time, and a given segment of a protein molecule will have different structures at different time points. As a result, at any given moment, an IDP has a structure which is different from a structure seeing at another moment [91].

Another level of structural heterogeneity is determined by the fact that many proteins are hybrids of ordered and disordered domains, and that this mosaic structural organization is crucial for their functions [13]. Also, even when IDPs do not possess ordered domains, they are known to have various levels and depths of disorder [52]. Over the past few years, an understanding of the available conformational space of IDPs/IDPRs underwent significant evolution. In fact, for a long time, IDPs were considered mostly 'unstructured' or natively 'unfolded' polypeptide chains. This was mostly due to the fact that the majority of IDPs analyzed at early stages of the field contained very little ordered structure; i.e., they were really mostly unstructured or unfolded. The discovery and characterization of such 'structure-less' proteins was important to build up a strong case to counter-point the dominant view represented by the classical sequence-to-structure-to-function paradigm, especially since such fully unstructured, yet functional proteins clearly represented the other extreme of the protein structure-function spectrum [13]. Top half of the Fig. 4 illustrates this situation by opposing rock-like ordered proteins and cooked spaghetti-like IDPs.

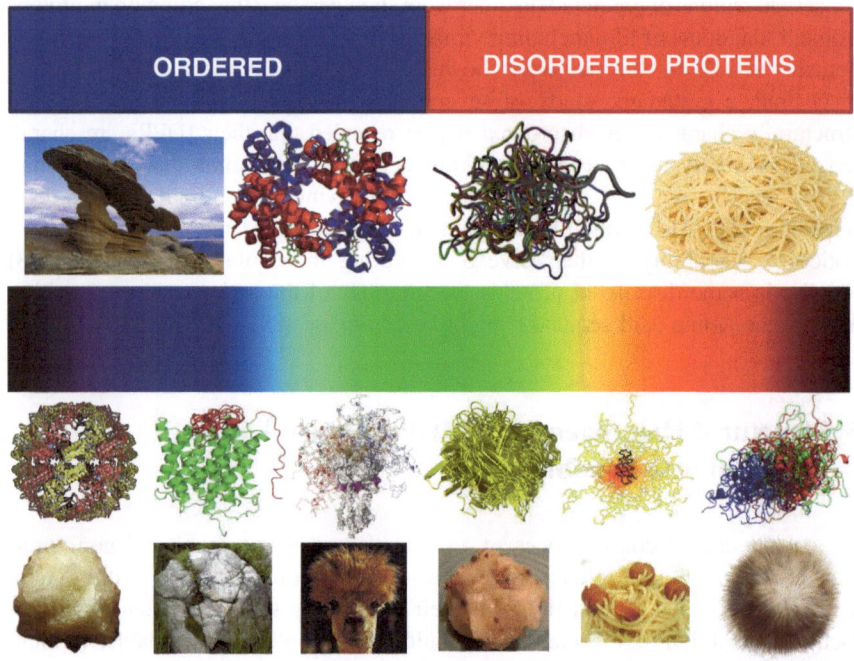

Fig. 4 Structural heterogeneity of IDPs/IDPRs. *Top half* Bi-colored view of functional proteins which are considered to be either ordered (*folded, blue*) or completely structure-less (disordered, *red*). Ordered proteins are taken as rigid rocks, whereas IDPs are considered as completely structure-less entities, kind of cooked noodles. *Bottom half* A continuous emission spectrum representing the fact that functional proteins can extend from fully ordered to completely structure-less proteins, with everything in between. Intrinsic disorder can have multiple faces, can affect different levels of protein structural organization, and whole proteins, or various protein regions can be disordered to a different degree. Some illustrative examples includes ordered proteins that are completely devoid of disordered regions (*rock-like* type), ordered proteins with limited number of disordered regions (*grass-on-the rock* type), ordered proteins with significant amount of disordered regions (*lhama/camel hair* type), molten globule-like collapsed IDPs (*greasy ball* type), pre-molten globule-like extended IDPs (*spaghetti-and-sausage* type), and unstructured extended IDPs (*hairball* type). This figure is reproduced from Ref. [5]

However, already in some early studies, it was evident that IDPs/IDPRs could be crudely grouped into two major structural classes, proteins/regions either with compact or extended disorder [3, 4, 9, 44, 92]. Based on these observations, protein functionality was ascribed to at least three major protein conformational states: ordered, molten globular, and coil-like [10, 92], indicating that functional IDPs can be less or more compact and possess smaller or larger amounts of flexible secondary/ tertiary structure [3, 4, 9, 10, 44, 92]. Roughly at the same time, it was emphasized that extended IDPs (known at that time as natively unfolded proteins) do not represent a uniform group but consist of two broad structural classes: native coils and native pre-molten globules [3]. Currently available data suggest that intrinsic disorder possesses multiple flavors, can have multiple faces, and can affect different levels of protein structural organization, where whole proteins, or various protein regions can be disordered to different degrees [91]. This new view of conformational space of functional proteins can be visualized in a form a continuous spectrum of differently disordered species extending from fully ordered to completely structureless proteins, with a broad gradient between (Fig. 4, bottom half).

Therefore, functional proteins can be well-folded and be completely devoid of disordered regions (rock-like scenario). Other functional proteins may contain limited number of disordered regions (a grass-on-the rock scenario), or have significant amount of disordered regions (a llama/camel hair scenario), or be molten globule-like (a greasy ball scenario), or behave as pre-molten globules (a spaghetti-and-sausage scenario), or be mostly unstructured (a hairball scenario). Notably, in this representation, there is no boundary between ordered proteins and IDPs, and, the structure-disorder space of a protein is considered as a continuum. It is important to remember that even the most ordered proteins do not resemble "solid rocks" and have some degree of flexibility. In fact, a protein molecule is an inherently flexible entity and the presence of this flexibility (even for the most ordered proteins) is crucial for its biological activity [93]. Also, another important point to remember is that due to their heteropolymer nature, proteins are never random coils and always have some residual structure [91].

6 Typical Functions of IDPs and IDPRs

High natural abundance of IDPs/IDPRs and their specific structural features indicate that these proteins and regions carry out important biological functions. The importance of the lack of ordered structure in function of IDPs/IDPRs has been confirmed by several comprehensive studies [3, 8–11, 44, 51, 52, 94–99]. Furthermore, sites of posttranslational modifications (acetylation, hydroxylation, ubiquitination, methylation, phosphorylation, etc.) and proteolytic attack are frequently associated with regions of intrinsic disorder [10].

In agreement with this hypothesis, more than 150 proteins have been recently identified as containing functional disordered regions, or being completely disordered, yet performing vital cellular roles [10, 95]. Twenty-eight separate functions

Table 1 Functional classes of disorder

Class	Example	Function
Entropic chains	Microtubule-associated protein 2	Entropic bristle, spacing in microtubule architecture
Effectors	4EBP1, 2, 3	Inhibitor of translation initiation
Scavengers	Caseins	Inhibition of calcium precipitation in milk
Assemblers	Caldesmon	Actin polymerization
Display sites	CREB transactivator domain	Regulation by phosphorylation

Based on data reported in Ref. [11]

were assigned for these disordered regions, including molecular recognition via binding to other proteins, or to nucleic acids [95, 96]. An alternative view is that functional disorder fits into at least five broad classes based on their mode of action [11]. These broad classes, with illustrative examples, are outlined in Table 1. Recent research added chaperones to this list [11].

A broad spectrum of biological functions associated with IDPs/IDPRs was found based on a computational study carried out for the evaluation of a correlation between the functional annotations in the Swiss-Prot database and the predicted intrinsic disorder [100–102]. The approach is based on the hypothesis that if a function described by a given keyword relies on intrinsic disorder, then the keyword-associated protein would be expected to have a greater level of predicted disorder compared to the protein randomly chosen from the Swiss-Prot. To test this hypothesis, functional keywords associated with 20 or more proteins in Swiss-Prot were found and corresponding keyword-associated datasets of proteins were assembled. For each keyword-associated set, one thousand length-matching and number-matching sets of random proteins were drawn from Swiss Prot. Order-disorder predictions were carried out for the keyword-associated sets and for the matching random sets. If a function described by a given keyword were carried out by a long region of disordered protein, one would expect the keyword-associated set to have a greater amount of predicted disorder compared to the matching random sets. The keyword-associated set would be expected to have less prediction of disorder compared to the random sets if the keyword-associated function were carried out by structured protein. Given the predictions for the function-associated and matching random sets, it is possible to calculate p-values, where a p-value >0.95 was used to define a disorder-associated function and a p-value <0.05 was used to define an order-associated function. Intermediate p-values are ambiguous [100–102]. The application of this approach revealed that out of 710 Swiss-Prot keywords, 310 functional keywords were associated with ordered proteins, 238 functional keywords were attributed to disordered proteins, and the remainder 162 keywords yield ambiguity in the likely function-structure associations [100–102]. Interestingly, in that series of papers, most of the structured protein-associated key words were shown to be related to enzymatic activities, whereas the majority of disordered protein-associated keywords were related to signaling and regulation. These results agree well with the notion that enzymatic catalysis requires ordered structure whereas the effectiveness of signaling is more dependent on binding reversibility, a property directly associated with the thermodynamics of disorder-to-order transition induced by binding.

IDPs/IDPRs are abundantly involved in numerous biological processes [3, 4, 8–12, 32, 44, 51, 52, 65, 91, 92, 95–98, 100–117], where they are found to play different roles in regulation of the functions of their binding partners and in promotion of the assembly of supra-molecular complexes. The conformational plasticity associated with intrinsic disorder provides IDPs/IDPRs with a wide spectrum of exceptional functional advantages over the functional modes of ordered proteins and domains [4, 8, 10, 51, 52, 95–98, 106–108, 110]. For example, the high accessibility of sites within the disordered proteins simplifies their post-translational modifications, such as phosphorylation, acetylation, lipidation, ubiquitination, sumoylation, etc., allowing for a simple mean of the modulation of their biological functions [4]. Many IDPRs contain specific identification regions via which they are involved in various regulation, recognition, signaling and control pathways [51, 52]. Conformational plasticity confers numerous advantages to the intrinsic disorder-based protein interactions [7, 8, 10, 33, 73, 106, 118].

Obviously, for these proteins, the predominant structure-function paradigm is insufficient, which suggests a more comprehensive view of the protein structure/function relationships is needed. In fact, a new paradigm was offered [10, 92, 95] to elaborate the sequence-to-structure-to-function scheme in a way that includes the novel functions of disordered proteins (Fig. 5). The complex data supporting this revised view were summarized in the Protein Trinity Hypothesis [92], which

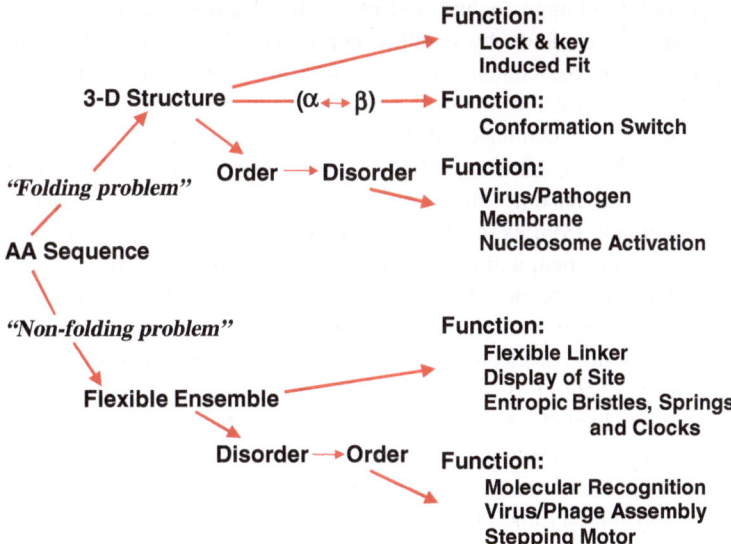

Fig. 5 Involvement of intrinsic disorder in protein function. Note that the classical structure-function paradigm cannot describe many of the function performed by proteins

suggested that native proteins can be in one of three states, the solid-like ordered state, the liquid-like collapsed-disordered state or the gas-like extended-disordered state. Function is then viewed to arise from any one of the three states or from transitions between them. This model was subsequently expanded to include a fourth state (pre-molten globule) and transitions between all four states [3].

It is necessary to note that some important activities related to molecular recognition do not directly involve coupled binding and folding, but rather are dependent on the flexibility, pliability and plasticity of the backbone. We are calling these "entropic chain activities", as they rely entirely on an extended random-coil conformation of a polypeptide that has to maintain constant motion during functioning. Discovering new entropic chain activities and estimating the fraction of ID regions involved in such activities are both intriguing problems [10]. One illustrated example of such entropic chain activities is considered below. Voltage gated ion channels cycle among three states: closed (sensitive to voltage), open, and inactive (insensitive to voltage). In the ball and chain mechanism for ion channel inactivation, a highly flexible "chain" carries out a random search until the "ball" plugs the open channel (Fig. 6). Portions of this figure are based on findings of Antz et al. [119] For simplicity, only four of the proposed ten states are shown [120]. The inactivation depends on a binding interaction between the channel opening and the "ball". The time of opening is also a crucially important. This "time of being open" function depends directly on the length and flexibility of the disordered "chain." An extended disordered region functions as one component of an entropic clock found in some ion channels. Charge migrations within the tetrameric pore proteins are associated with the majority of state changes of voltage-gated K^+ ion channels [120]. However, the timing of the inactivation step is determined by the time it takes for a mobile domain to find and block the channel. The movement of the mobile domain is restricted by a tether composed of ~ 60 disordered residues (Fig. 6). The timing of channel inactivation is a function of the length of the disordered tether. Since ion channels serve to modulate the excitability of nerve cells, their malfunction can have substantial impact on human health. Mutations in the human homolog of the Shaker K^+ channel (KCNA1) can lead to myokymia, partial epilepsy, or episodic ataxia [121, 122]. The ball and chain model was originally developed from experiments showing that protease digestion caused the open state to remain open, and then adding back the trypsin-released peptide led to channel inactivation. Recent NMR data provides direct confirmation of the flexibility of the "chain" region. If channel closure were to involve a random search by a flexible chain, the time of closure would be inversely proportional to the square of the length of the chain [123]. Genetic engineering of the Shaker K^+ channel was used to construct channels chains of various lengths. The inactivation times for the chains followed the expected dependence on length. Shorter chains speeded up inactivation and longer chains slowed inactivation [124]. Taken together these data strongly support the ball and chain mechanism.

It has been pointed out that in a living organism proteins participate in complex interactions, which represent the mechanistic foundation of the organism's physiology and function. Regulation, recognition and cell signaling involve the

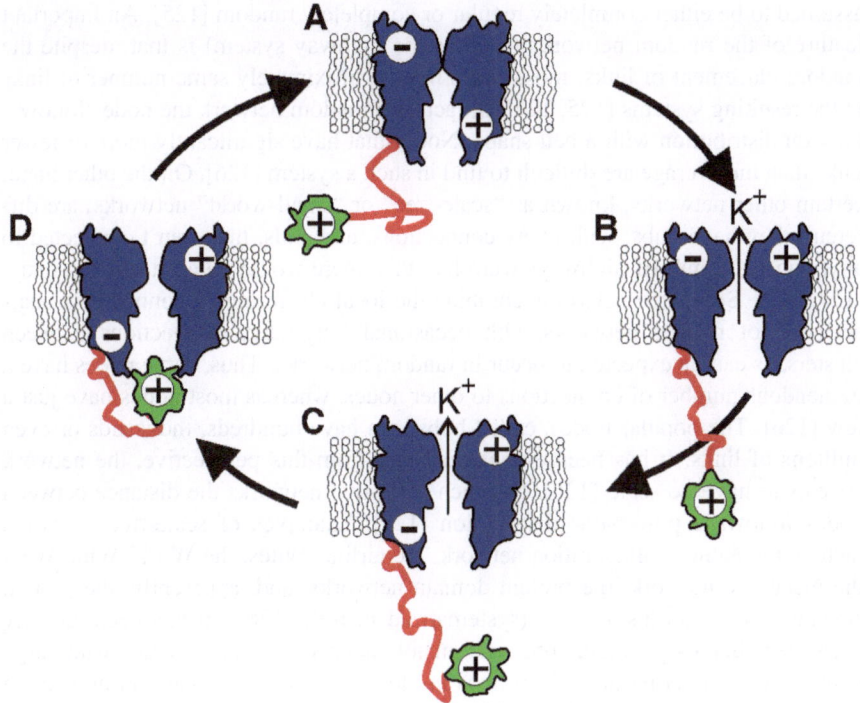

Fig. 6 Example of an entropic clock: Simplified model of a Shaker-type voltage-gated K$^+$ ion channel (*blue*) with 'ball and chain' timing mechanism. The 'ball and chain' is comprised of an inactivation, or ball, domain (*yellow*) that is tethered to the pore assembly by a disordered chain (*red*) of ~60 residues. For simplicity, only four of the proposed ten states are shown. The cytoplasmic side of the assembly is oriented downward. *A* Closed state prior to membrane depolarization. Note that conformational changes of the pore have sealed the channel and a positive charge on the cytoplasmic side of the pore assembly excludes binding of the ball domain. *B* Open state following membrane depolarization. *C* After depolarization, the cytoplasmic side of the pore opening assumes a negative charge that facilitates interaction with the positively charged ball domain. *D* Inactivation of the channel occurs when the ball domain occludes the pore. The transition from *C* to *D* does not involve charge migration and can be modeled as a random walk of the ball domain towards the pore opening. Portions of figure are based on findings in Antz et al. [119]

coordinated actions of many players. To achieve this coordination, each participant must have a valid identification (ID) that is easily recognized by the others. For proteins, these IDs are often within IDPRs [51, 52]. Thus, intrinsically disordered regions are typically involved in regulation, recognition, signaling and control pathways in which interactions with multiple partners and high-specificity/low-affinity interactions are often requisite. In this way, the functional diversity provided by disordered regions complements those of ordered protein regions.

IDPs as hubs in protein signaling networks It is known that many diverse systems may best be described as networks with complex topologies, which is often assumed to be either completely regular or completely random [125]. An important feature of the random networks (such as the highway system) is that, despite the random placement of links, most nodes have approximately same number of links in the resulting systems [125, 126]. In fact, in a random network the nodes follow a Poisson distribution with a bell shape. Nodes that have significantly more or fewer links than the average are difficult to find in such a system [126]. On the other hand, certain other networks, known as "scale-free" or "small-world" networks, are different: they have hubs, with many connections, and ends, that aren't connected to anything but a hub (if highways were like this, there would be a lot of dead-ends) [127, 128]. Scale-free networks combine the local clustering of connections characteristic of regular networks with occasional long-range connections between clusters, as can be expected to occur in random networks. Thus, some nodes have a tremendous number of connections to other nodes, whereas most nodes have just a few [126]. The popular nodes, called hubs, can have hundreds, thousands or even millions of links. It has been emphasized that from this perspective, the network appears to have no scale [126]. For such scale-free networks the distance between nodes follows a power-law distribution [129]. Examples of scale-free networks include the author-collaboration network, the airline routes, the World Wide Web, the metabolic network, the protein domain networks, and, apparently, the protein signaling networks inside cells (systematized in Ref. [128]). It has been recently suggested that α-synuclein, p53 and many other IDPs that interact with large numbers of distinct partners form hubs in the scale-free protein-protein interaction network inferred for the cell [51, 52].

Scaffold proteins Scaffold proteins constitute an important subclass of hubs that typically have a modest number of interacting partners and that are commonly found at the central parts of functional complexes, where they interact with most of their partners at the same time and therefore act as party hubs [130]. Besides being responsible for bringing together specific proteins within a signaling pathway and providing selective spatial orientation and temporal coordination to facilitate and promote interactions among interacting proteins, some scaffolds can influence the specificity and kinetics of signaling interactions via simultaneous binding to multiple participants in a particular pathway and facilitation and/or modifying the specificity of pathway interactions [131], other scaffold can change conformations of individual proteins and thus modulate their activities [131], still other scaffold proteins may modulate the activation of alternative pathways by promoting interactions between various signaling proteins [106]. Analysis of several well-characterized signaling scaffold proteins reveled that their large IDPRs are crucial for the successful scaffold function [106]. A more global bioinformatics analysis of 74 scaffold proteins revealed that a typical design of a scaffold protein includes a set of short globular domains (~ 80 amino acids on average) connected by long linker regions (~ 150 residues on average) with crucial binding functions [87]. This gave further support to the notion that signaling scaffold proteins utilize the various features of highly flexible ID regions to obtain more functionality from less structure [106].

Disorder and transcription regulation Conformational plasticity and adaptability associated with intrinsic disorder are crucial for various protein functions. Among the proteins whose functional life is strongly disorder-dependent are transcription factors (TFs) [132, 133] and other proteins involved in transcriptional regulation, such as the mediator complex [26, 134], core and linker histones [135], and ribosomal proteins [136]. For example, based on the comprehensive computational analysis it was concluded that from 83 to 94 % of TFs might possess long IDPRs, with the degree of disorder in eukaryotic TFs being significantly higher than in prokaryotic TFs. Also, TFs were shown to be substantially depleted in order-promoting residues and significantly enriched in disorder-promoting residues, and were characterized by high levels of α-MoRF (molecular recognition feature) [132]. Furthermore, the analysis of the distribution of disorder within the TFs revealed that the degree of disorder in their activation regions is much higher than that in DNA-binding domains and that the AT-hooks and basic regions of TF DNA-binding domains are highly disordered suggesting that eukaryotes with their well-developed gene transcription machinery require transcription factor flexibility to be more efficient [132]. A number of interesting and important roles were also ascribed to intrinsic disorder in TFs related to the regulation of heat shock response (so called heat shock factors, HSFs) [137] and in the reprogramming transcription factors (the so called Yamanaka factors, namely Sox2, Oct3/4 (Pou5f1), Klf4, and c-Myc, and the Thomson factors, namely Sox2, Oct3, Lin28, and Nanog) overexpression of which is known to generate induced pluripotent stem (iPS) cells from terminally differentiated somatic cells [138].

Disorder in regulation of cellular pathways Of special interests are the vital roles of intrinsic disorder in regulation and orchestration of various cellular pathways. One of the illustrative examples of this regulatory role of intrinsic disorder is the canonical Wnt-pathway that involves five proteins, Axin, CKI-α, GSK-3β, APC, and β-catenin (all shown to contain long IDPRs), that is known to play a number of crucial roles in the development of organism, and the malfunctions of which might lead to various diseases including cancer [139]. Here, based on the comprehensive analysis of published data it was concluded that IDPRs found in Wnt-pathway proteins orchestrate protein-protein interactions, and facilitate post-translational modifications and signaling. Furthermore, the scaffold protein Axin and another large protein, APC, are heavily enriched in disorder and act as flexible concentrators in gathering together all other proteins involved in the Wnt-pathway. Intriguingly, the multifarious roles of highly disordered APC in regulation of β-catenin function were established by showing that disordered APC helps the collection of β-catenin from cytoplasm, facilitates the β-catenin delivery to the binding sites on Axin, and controls the final detachment of β-catenin from Axin [139].

Another important illustration of the involvement of intrinsic disorder in regulation of crucial pathway is given by the process of the programmed cell death (PCD), which is one of the most intricate cellular processes where the cell uses specialized cellular machinery and intracellular programs to kill itself and which enables metazoans to control cell numbers and eliminate cells that threaten the animal's survival [140]. PCD includes several specific modules, such as apoptosis, autophagy and programmed necrosis (necroptosis). These modules are not only tightly regulated but also intimately interconnected and are jointly controlled via a complex set of protein-protein interactions. Recently, several large sets of PCD-related proteins across 28 species were analyzed using a wide array of modern bioinformatics tools to understand the role of the intrinsic disorder in controlling and regulating the PCD [140]. This analysis revealed that proteins involved in regulation and execution of PCD possess substantial amount of intrinsic disorder and IDPRs were implemented in a number of crucial functions, such as protein–protein interactions, interactions with other partners including nucleic acids and other ligands, were shown to be enriched in posttranslational modification sites, and were characterized by specific evolutionary patterns [140].

Disorder-based competitive recruitment Analysis of the peculiarities of interaction between the intrinsically disordered antibiotic protein colicin E9 (ColE9) with the Tol-Pal system revealed that intrinsic disorder plays crucial role in promoting successful competition of IDP with other proteins for binding partners [141]. The Tol-Pal system spans the cell envelope of Gram-negative bacteria and contributes to the stability and integrity of the bacterial outer membrane (OM). It consists of seven proteins: TolA, TolQ and TolR constitute an inner membrane (IM) complex, a peptidoglycan-associated lipoprotein Pal is anchored in the OM, YbgC is found in the cytoplasm, and TolB and YbgF are periplasmic proteins. A soluble periplasmic protein TolB interacts with Pal forming the OM complex which is tightly associated with structural components of the cell envelope. Furthermore, TolB interacts with TolA therefore linking the IM and OM complexes [141].

The output of the ColE9 interaction with the Tol-Pal system has deadly consequence. In fact, ColE9 hijacks the Tol-Pal system to translocate its cytotoxic endonuclease (DNase) domain across the OM. Then, this domain randomly degrades the bacterial genome, eventually leading to the cell death. ColE9 is known to be a very effective killer, and a single colicin molecule is believed to be sufficient to kill a cell. At the heart of ColE9-mediated road to kill is the competitive interaction of ColE9 with TolB at the specific canyon in the TolB β-propeller domain that overlaps with the binding site of Pal [142]. Intrinsically disordered TolB binding epitope of ColE9 (TBE, residues 32–47) is a part of the N-terminally located 83-residue Intrinsically Unstructured Translocation Domain (IUTD) [142]. Although ColE9 TBE interacts with the Pal-binding cite of TolB and mimics some of the Pal interactions, a set of specific interactions is also formed determining the different allosteric response from TolB [142]. As a result, interaction of TolB with TolA is promoted triggering the toxin translocation through the OM [143].

Using a series of pre-steady-state kinetic experiments it was shown that ColE9 is able to compete very efficiently with the ordered Pal for binding to TolB [141]. This analysis also revealed the kinetic basis of the competitive TolB recruitment by the intrinsically disordered ColE9 TBE and showed that the efficiency of this recruitment is strongly Ca^{2+}-dependent and is determined by the combination of rapid binding and slow dissociation. Here, disordered ColE9 was shown to first rapidly bind to TolB through a simple bimolecular mechanism. Binding rate was not affected by the Ca^{2+} ions, but the rate of complex dissociation was and was shown to decrease in the presence of Ca^{2+} due to the increase in the ColE9-TolB complex stability. Contrarily to this, both formation and dissociation of Pal-TolB complex were slow calcium-independent processes. Another intrigue is in the structural outcomes of the ColE9 and Pal binding to TolB, at ColE9-TolB complex formation, intrinsically disordered ColE9 folds, but structure of ordered TolB does not change, whereas in Pal-TolB complex, ordered Pal remains almost unchanged, but TolB undergoes noticeable structural rearrangements [143]. This study clearly revealed that even under the equimolar concentrations of ColE9, TolB, and Pal, intrinsically disordered ColE9 TBE is able to efficiently recruit TolB.

Importantly, the process of the TolB-ColE9 TBE complex formation includes both the binding-induced folding of the intrinsically disordered ColE and the disruption of the non-native hydrophobic clusters seeing in the unbound ColE9 [144], a step preceding the epitope association. Therefore, the competitive recruitment of TolB by ColE9 was proposed to start with the disruption of the non-native contacts in the functionally misfolded [145] ColE9 likely in its encounter complex with TolB, followed by the fast but weak binding of the unfolded ColE9 to the TolB β-propeller tunnel, followed by the structural rearrangement of ColE9 and the recruitment of Ca^{2+} ions, eventually leading to the locking of ColE9 within the TolB binding site canyon [141]. Importantly for successful hijacking schema, all these events occur faster than the extensive conformational changes in the TolB-Pal complex [141].

Functional regulation via alternative splicing Alternative splicing (AS) is a process by which two or more mature mRNAs are produced from a single precursor pre-mRNA by the inclusion and omission of different segments [146, 147]. AS is commonly observed mostly in multicellular eukaryotes [148]. For humans and other mammals, multiple proteins are often produced from a single gene since 40–60 % the genes yield proteins via the AS mechanism [149–151]. It was hypothesized that AS very likely provides an important mechanism for enhancing protein diversity in multicellular eukaryotes [152]. AS has effects on a diversity of protein functions such as protein-protein interactions, ligand binding, and enzymatic activity [153–155]. Removal of a piece of sequence from within a structured protein would often lead to protein misfolding and cause protein aggregation and loss of function. Therefore, in structured proteins, AS-induced alterations are generally of small size, are usually located on the protein surface, and are most often located in coil regions [156]. Given the small sizes and locations of the changes resulting from alternative splicing, the different splice variants were predicted to fold into the same overall structures, with only slight structural perturbations that could be functionally important [156, 157]. On the other hand, when AS was to map to ID regions, both multiple and long splice variants were allowed because structural perturbation would not be a problem [54]. These findings have crucial functional implementations. Since disorder plays various roles in protein functions and in protein-protein interaction networks, modification of such functions could be readily accomplished by AS within disordered regions.

Figure 7 illustrates this idea by showing how AS modifies the functional repertoire of an important tumor suppressor protein BRCA-1, which is known to

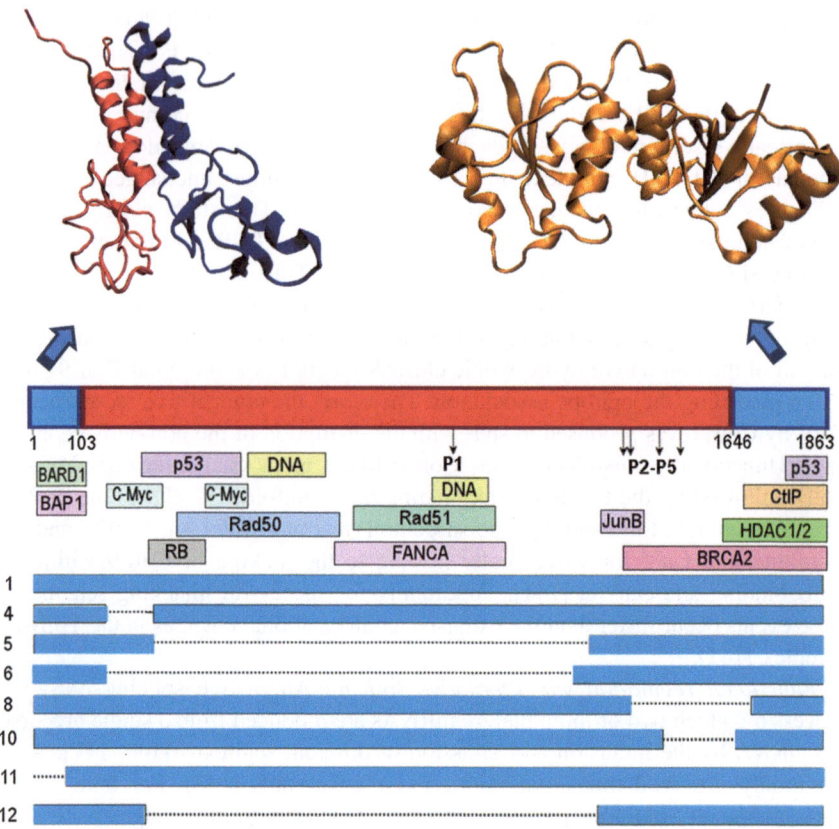

Fig. 7 Modulating protein function with alternative splicing. Functional domains of BRCA1 and representative alternatively spliced isoforms [54]. The *blue cylinders* correspond to the N-terminal and C-terminal ordered domains, whereas a central IDPR of ∼1,500 residues is shown as *red cylinder* at the *middle* of BRCA1. The *vertical arrows* show the location of phosphorylation sites, and the *colored rectangles* represent binding domains, including the binding partner name. Isoforms are shown at the *bottom*, with *dotted lines* representing the regions missing from the translated protein products due to alternative splicing. The numbers at the *left* are the isoform identifiers. The missing numbers (*2, 3, 7,* and *9*) correspond to the BRCA1 isoforms for which there is no structural information. After Ref. [54]. Structures of ordered domains of BRCA1 are shown on the *top panels* of this figure: N-terminal RING-type domain (residues 1–110, *red structure*) in complex with the RING domain of BRCA1-associated ring domain protein 1 (*blue structure*), PDB ID: 1JM7; C-terminal BRCT1-BRCT2 domains (residues 1,649–1,859, *orange structure*), PDB ID: 1T15

participates in many cellular pathways, such as transcription, apoptosis, and DNA repair, through direct or indirect interaction with a variety of partners [158]. A canonical isoform of this protein is a 1,863 residues-long polypeptide that contain a long intrinsically disordered central region [159], flanked by ordered domains at the two termini, an N-terminal RING finger domain of 103 residues, and C-terminus located two tandem copies of the BRCA1 C-terminal domain with a total of 218 residues making up the two domains.

Figure 7 shows that the disordered central region of BRCA1 contains molecular recognition domains for both DNA and several protein-binding partners. It also shows that different AS isoforms described for BRCA1 lack various functional regions [54]. This lack of specific functional regions in various AS isoforms creates diverse functional profiles for the transcribed gene products. Furthermore, it was emphasized that since regulatory and signaling elements in disordered regions can be comprised of just a few more or less continuous amino acids, and since a high density of functionally important segments can be located within the IDPRs, functionality of IDPs can be completely rewired via the AS [54]. Thus, a linkage between AS and signaling by disordered regions provides a novel and plausible mechanism for understanding the origins of cell differentiation, which ultimately gave rise to multicellular organisms in nature [54].

Functional regulation via posttranslational modifications Posttranslational modifications (PTMs) of proteins are reversible or irreversible chemical changes of a polypeptide chain that occur after DNA has been transcribed into RNA and translated into protein. PTMs range from the enzymatic cleavage of peptide bonds to the covalent additions of particular chemical groups, lipids, carbohydrates or even entire proteins to amino acid side chains. PTMs extend the range of amino acid structures and properties, and diversify structures and functions of proteins [160]. Although DNA typically encodes 20 primary amino acids, proteins contain more than 140 different residues, because of various PTMs. In general, proteomes are significantly more complex than one can expect from the analysis of the encoding genomes due mostly to two mechanisms, PTMs and alternative splicing of their mRNAs (see below) [160]. Some PTMs (e.g., phosphorylation) are readily reversible by the action of specific deconjugating enzymes. The interplay between modifying and demodifying enzymes allows for rapid and economical control of protein function. A similar control by protein degradation and de novo synthesis would take much longer time and cost much more energy.

PTM can occur at any stage of the protein life. Some proteins are modified shortly after their translation is completed and prior to the final steps of their folding. These early PTMs might affect the protein folding efficiency, protein conformational stability and even determine the fate of the nascent protein via directing it to distinct cellular compartments. Other proteins are modified after their folding and localization are completed. Here, PTMs can activate or inactivate catalytic activity or otherwise influence the biological activity of a protein.

PTMs come in a wide variety of types, and are mostly catalyzed by special enzymes that recognize specific target sequences in specific proteins. In higher eukaryotes, as much as 5 % of the genomes are expected to encode enzymes related to the posttranslational modifications of the proteomes. Altogether, as many as 300 post-translational modifications of proteins are known to occur physiologically [161]. The most common PTMs are the specific cleavage of precursor proteins; formation of disulfide bonds; or covalent addition or removal of low-molecular-weight groups, thus leading to modifications such as acetylation, amidation, biotinylation, cysteinylation, deamidation, farnesylation, formylation, geranylgeranylation, glutathionylation, glycation (nonenzymatic conjugation with carbohydrates), glycosylation (enzymatic conjugation with carbohydrates), hydroxylation, methylation, mono-ADP-ribosylation, myristoylation, oxidation, palmitoylation, phosphorylation, poly(ADP-ribosyl)ation, stearoylation, or sulfation [162]. All amino acid side chains are known to undergo chemical diversification due to various PTMs. However, most often protein PTMs are found at side chains that can act as either strong (C, M, S, T, Y, K, H, R, D, E) or weak (N, Q) nucleophiles, whereas the remaining residues (P, G, L, I, V, A, W, F) are rarely involved in covalent modifications of their side chains [160]. Table 2 lists major PTMs described for various amino acid side chains and also shows some illustrative examples.

PTMs play a number of fundamental roles in regulating the folding of proteins, their targeting to specific subcellular compartments, their interaction with ligands or other proteins, and their functional state, such as catalytic activity in the case of enzymes or the signaling function of proteins involved in signal transduction pathways [163, 164]. Some proteins require multiple different types of posttranslational modifications for their function. For such multi-PTM proteins, modified sites in proteins can not only mediate individual functions, but can also function together to fine-tune molecular interactions and to modulate overall protein activity and stability [165]. One dramatic example of such proteins is provided by a family of nuclear intrinsically disordered proteins, histones, which require methylation, acetylation, phosphorylation, ubiquitylation, ADP-ribosyation, and SUMOylation at different stages of their action, with different modifications affecting histone–DNA interactions and also the histone–histone interfaces, thus providing the capacity to disrupt intra-nucleosomal interactions and to alter nucleosome stability [135]. Although the N-terminal domains of the core histones are known to contain an extraordinary number of sites that can be subjected to PTM, over 30 histone modifications have been recently identified in the core domains too [166].

Typically, PTMs are classified according to the involved mechanisms: addition of functional groups (e.g., acylation, alkylation, phosphorylation, glucosylation, etc.); attachment of other proteins and peptides (e.g., ubiquitination, SUMOylation, etc.); changing of the chemical nature of amino acids (deamidation, deimidation, oxidation, etc.); and dissection of the backbone by proteolytic cleavage [102]. Another type of PTM classification involves description of the fragment of cosubstrate or coenzyme that is enzymatically coupled to the protein and the chemical nature of the protein modification [160]. This includes S-adenosylmethionine (SAM)-dependent methylation, phosphoadenosinephosphosulfate (PAPS)-dependent sulfurylation, NAD-dependent ADP ribosylation, acetyl CoA dependent acetylation, ATP-dependent phosphorylation, and CoASH-dependent phosphopantetheinylation. PTMs can also be classified via consideration of the new function enabled by the covalent addition. These include gain in catalytic function (seeing in enzymes that have acquired tethered biotinyl, lipoyl, and phosphopantetheinyl groups), changes of subcellular address for proteins undergoing various lipid modifications (prenylation, palmitoylation, glycosyl phosphatidylinositol (GPI) anchor attachment), and targeting of the modified proteins for proteolytic destruction by ubiquitylation to mark transport to lysosomes or proteasomes.

Table 2 Variability of post translational modifications of the side chains in proteins

Residue	Reaction	Example
Alanine (Ala, A)	N-acetylation	N-alpha-acetyltransferase
	Amidation	Pantothenate synthetase
	N-methylation	Ribosomal proteins
Arginine (Arg, R)	N-methylation	Histones
	N-ADP-ribosylation	GSa
	Citrullination/Deimination	Argininosuccinate synthase
	N-acetylation	N-alpha-acetyltransferase
	Amidation	Tachykinins
	Dihydroxylation	Steilins
	Hydroxylation	Carbon monoxide dehydrogenase large chain
	Phosphorylation	Histones
	N-glycosylation	N-glycoproteins
Asparagine (Asn, N)	N-glycosylation	N-glycoproteins
	N-ADP-ribosylation	eEF-2
	Protein splicing	Intein excision step
	Deamidation	Isomerization to isoaspartate and aspartate
	Amidation	FMRFamide-related peptides
	Hydroxylation	Hypoxia-inducible factor 1-alpha
Aspartic acid (Asp, D)	Phosphorylation	Protein tyrosine phosphatases; response regulators in two component systems
	Isomerization to isoaspartate	Protein-L-isoaspartate O-methyltransferase
	Deamidation	Beta-casein
	N-acetylation	N-alpha-acetyltransferase
	Beta-methylthiolation	Ribosomal proteins
	Hydroxylation	3-hydroxyaspartate alsolase
	Cis-14-hydroxy-10,13-dioxo-7-heptadecenoic acid aspartate ester	Non-specific lipid transfer proteins
Cysteine (Cys, C)	S-hydroxylation (S-OH)	Sulfenate intermediates; peroxiredoxins
	Disulfide bond formation	Protein in oxidizing environments
	Phosphorylation	PTPases
	S-acylation	Ras
	S-prenylation	Ras
	Protein splicing	Intein excisions
	N-acetylation	N-alpha-acetyltransferase
	N-ADP-ribosylation	Glyceraldehyde-3-phosphate dehydrogenase
	Amidation	Cystein synthase A
	S-archaeol cysteine	Halocyanin

(continued)

Table 2 (continued)

Residue	Reaction	Example
	Cysteine sulfinic acid (-SO₂H)	Cysteine sulfinic acid decarboxylase
	Methylation	Crystallins
	N-myristoylation	Genome polyproteins of several viruses
	Nitrosylation	Thioredoxins
	N-palmitoylation	Small cystein-rich outer membrane protein
	S-palmitoylation	OmcA
	S-glutathionylation	Myelin proteolipid proteins Redox regulation via reversible glutathionylation
Glutamic acid (Glu, E)	Methylation	Chemotaxis receptor proteins
	Carboxylation	Gla residues in blood coagulation
	Polyglycination	Tubulin
	Polyglutamylation	Tubulin
	N-acetylation	N-alpha-acetyltransferase
	Poly N-ADP-ribosylation	Poly [ADP-ribose] polymerase 1
	Amidation	Buccalin
	Deamidation followed by methylation	Methyl-accepting chemotaxis proteins
Glutamine (Gln, Q)	Transglutamination	Protein cross-linking
	Deamidation	Myelin basic protein
	Amidation	FMRFamide-related peptides
	N-methylation	Ribosomal proteins
Glycine (Gly, G)	C-hydroxylation	C-terminal amide formation
	N-acetylation	N-alpha-acetyltransferase
	Amidation	Glycine oxidase
	Cholesterol glycine ester	Hedgehog proteins
	N-myristoylation	Protein Nef
Histidine (His, H)	Phosphorylation	Sensor protein kinases in two-component regulatory systems
	Aminocarboxypropylation	Diphthamide formation
	N-methylation	Methyl CoM reductase
	Amidation	VIP peptides
	Bromination	Sperm-activated peptide SAP-b
	Methylation	Actin
Isoleucine (Ile, I)	Amidation	FMRFamide neuropeptides
	N-methylation	Fimbial protein
Leucine (Leu, L)	Amidation	Myomodulin neuropeptides
	N-methylation	Major structural subunit of bundle-forming pilus

(continued)

Table 2 (continued)

Residue	Reaction	Example
Lysine (Lys, K)	N-methylation	Histone methylation
	N-acylation by acetyl, biotinyl, lipoyl, ubiquityl groups	Histone acetylation; swinging-arm prosthetic groups; ubiquitin; SUMO (small ubiquitin-like modifier) tagging of proteins
	C-hydroxylation	Collagen maturation
	O-glycosylation	Adiponectin; O-glycoproteins
	N-acetylation	N-alpha-acetyltransferase
	Allysine	Elastin and collagen; lysyl oxidase
	Amidation	Histone-lysine N-methyltransferase EHMT1
	N6-1-carboxyethylation	Carbonyl reductases
	Dihydroxylation	Steilins
	Hydroxylation	Collagens
	N-myristoylation	Tumor necrosis factors
	N-palmitoylation	Serine palmitoyltransferases
	Trimethylation	Myosins
Methionine (Met, M)	Oxidation to methionine sulfoxide	Methionine sulfoxide reductase
	Oxidation to methionine sulfone	Catalase
	Silent modification (conversion to aspartic acid)	Unstable hemoglobin, Hb Bristol [p67 (E11) Val-Met]
	N-acetylation	N-alpha-acetyltransferase
	Amidation	MIP-related peptides
	N-methylation	Ribosomal proteins
Phenylalanine (Phe, F)	Amidation	FMRFamide neuropeptides
	Hydroxylation	Adhesive plaque matrix proteins
	N-methylation	ComG operon proteins
Proline (Pro, P)	C-hydroxylation	Collagen; HIF-1a
	Dihydroxylation	Virotoxin
	N-acetylation	N-alpha-acetyltransferase
	Amidation	Prothyroliberin
	N-methylation	N-methylproline demethylase
Serine (Ser, S)	Phosphorylation	Protein serine kinases and phosphatases
	O-glycosylation	Notch O-glycosylation
	Phosphopantetheinylation	Fatty acid synthase
	Autocleavages	Pyruvamidyl enzyme formation
	N-acetylation	N-alpha-acetyltransferase
	O-acetylation	O-acetylserine (thiol) liase

(continued)

Table 2 (continued)

Residue	Reaction	Example
	N-ADP-ribosylation	Ras-related protein Rap-1b
	Amidation	Kallikrein-8
	N-decanoylation	Ghrelin
	O-octanoylation	Appetite-regulating hormons; ghrelin
	O-palmitoylation	Myelin proteolipid protein
	Sulfation	Retrograde protein of 51 kDa
Threonine (Thr, T)	Phosphorylation	Protein threonine kinases/phosphatises
	O-glycosylation	O-glycoproteins
	N-acetylation	N-alpha-acetyltransferase
	Amidation	Aurora kinase A
	N-decanoylation	Ghrelin
	O-octanoylation	Ghrelin
	O-palmitoylation	Myelin proteolipid protein
	Sulfation	Cathepsin
	O-acetylation	Inhibitor of nuclear factor kappa-B kinase subunit alpha
Tryptophan (Trp, W)	C-mannosylation	Plasma-membrane proteins
	Amidation	Neuropeptide-like proteins
	Bromination	Mu-conotoxins
	C-linked glycosylation	Properdin
	Hydroxylation	Alpha-ketoglutarate-dependent taurine dioxygenase
Tyrosine (Tyr, Y)	Phosphorylation	Tyrosine kinases/phosphatases
	Sulfation	CCR5 receptor maturation
	ortho-nitration	Inflammatory responses
	TOPA quinine	Amine oxidase maturation
	N-Acetylation	N-alpha-acetyltransferase
	Amidation	FMRFamide-related neuropeptides
	N-methylation	General secretion pathway protein I
	O-glycosylation	S-layer protein SpaA
Valine (Val, C)	N-Acetylation	N-alpha-acetyltransferase
	Amidation	MIP-related peptides
	Hydroxylation	Conophans

In addition to the molecular mechanism-based classification of PTMs, the conformational state of the site where the modification would take place can also be used to group PTMs in two major classes. The first group involves modifications that are associated primarily with structured proteins and regions, whereas the second group combines modifications that are associated primarily with IDPs/ID-PRs [102]. The first PTM group (e.g. formylation, protein splicing, oxidation and covalent attachment of quinones and organic radicals) is crucial for providing moieties for catalytic functions, for modifying enzyme activities, and for stabilizing protein structure. The second PTM group relies on the low affinity, high specificity binding interactions between a specific enzyme and a substrate (a protein that has to be modified). Among this group of PTMs are phosphorylation, acetylation, acylation, adenylylation, ADP ribosylation, amidation, carboxylation, formylation, glycosylation, methylation, sulfation, prenylation, ubiquitination, and Ubl-conjugation (i.e., covalent attachment of ubiquitin-like proteins, including SUMO, ISG15, Nedd8, and Atg8). A particular advantage of disorder for regulatory and signaling regions is that changes, such as protein modification, lead to large-scale disorder-to-order structural transitions: such large-scale structural changes are not subtle and so could be an advantage for signaling and regulation as compared to the much smaller changes that would be expected from the decoration of an ordered protein structure.

The importance of intrinsic disorder for action of enzymes catalyzing different PTMs is well illustrated by kinases. It is estimated that functions of one-third of eukaryotic proteins are controlled via phosphorylation/dephosphorylation cycles that originate from carefully regulated protein kinase and phosphatase activities [167]. Phosphorylation sites are frequently located within functionally important protein domains. For example, the majority of phosphorylation sites of Mdm2 are located in its p53- and p14-ARF-binding regions, whereas ubiquitin-mediated degradation of many proteins is controlled by the phosphorylation of their PEST motifs. Often, Eukaryotic protein kinases constitute one of the largest gene families. For example, yeast kinome includes 119 kinases, there are 1,019 kinase- and 300 phosphatase-coding genes in *Arabidopsis thaliana*, there are 540 kinases in mouse kinome, whereas human genome contains ~ 520 genes encoding kinases and more than 150 genes encoding phosphatases. However, in any given proteome, the number of kinases and phosphatases is noticeably smaller than the number of their potential substrates. In fact, on average, each eukaryotic protein kinase serves ~ 20 substrates, whereas each human phosphatase is expected to dephosphorylate ~ 65 clients. This converts the classical "lock-and-key model" of the enzyme action into the "one-lock-many-keys" scenario. Although phosphorylation by each kinase is a highly specific process, kinase substrates typically bind to the enzyme with weak affinity. Such combination of high specificity and low affinity is characteristic for the intrinsic disorder-based signaling interactions. In fact, often such a combination of characteristics is achieved via coupled binding and folding. The low net affinity arises because the positive free energy associated with the disorder-to-order transition reduces the magnitude of the negative free energy arising from the interactions within the contact surface [168–173]. Figure 8 illustrates some of the peculiarities of such signaling interactions by presenting a crystal structure of the complex between a 20-amino acid peptide derived from the heat stable protein kinase inhibitor (PKIα) and the catalytic subunit of cyclic AMP-dependent protein kinase (cAPK, PDB ID: 2CPK). In this complex, PKIα is in a highly extended conformation which is stabilized by 36 H-bonds, among which there are only two intramolecular H-bonds, whereas 18 H-bonds are formed with water and remaining 16 H-bonds are formed with cAPK.

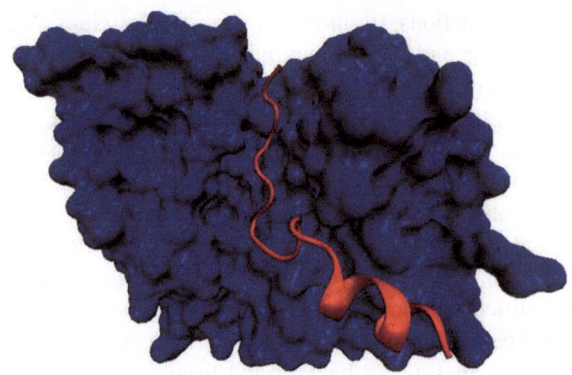

Fig. 8 Disordered nature of the PTM sites. Crystal structure of a complex between a 20-amino acid peptide derived from the heat stable protein kinase inhibitor (PKIα, *red chain*) and the catalytic subunit of cyclic AMP-dependent protein kinase (cAPK, *blue surface*). PDB ID: 2CPK

Bioinformatics analysis revealed that many sites of protein phosphorylation were found to be in regions structurally characterized as intrinsically disordered [95, 96]. In fact, amino acid compositions, sequence complexity, hydrophobicity, charge and other sequence attributes of regions adjacent to phosphorylation sites are very similar to those of intrinsically disordered protein regions [174]. Furthermore, there is a high correspondence between the prediction of disorder and the occurrence of phosphorylation [174].

In addition to phosphorylation, several other types of PTMs, such as acetylation, protease digestion, ubiquitination, fatty acid acylation, and methylation, have also been observed to occur in regions of intrinsic disorder [95, 96, 102, 175]. From these findings, it is tempting to suggest that sites of protein modification in eukaryotic cells universally or at least very commonly exhibit a preference for intrinsically disordered regions. For all of the examples discussed above, the modifying enzyme has to bind to and modify similar sites in a wide variety of proteins. If all the regions flanking these sites are disordered before binding to the modifying enzyme, it is easy to understand how a single enzyme could bind to and modify a wide variety of protein targets.

Some functional advantages of IDPs/IDPRs. Importantly, even sturdy locks (i.e., protein active sites) were shown to be rather flexible. In fact, as early as in 1958 it has been recognized that some compounds, being able to bind to the enzyme, were not processed to give a product. Some enzymes seemed to require smaller or larger degree of flexibility to be functional. To account for these discrepancies, a modification of the "lock and key" model called the "induced fit" theory has been proposed by Koshland [176]. According to this theory and its subsequent modifications/interpretations, the enzyme is partially flexible and the substrate does not simply bind to the active site, but it has to bring about changes to the shape of the active site to activate the enzyme and make the reaction possible. Substantial experimental evidence has been accumulated to support this view. For example, the existence of functional flexibility within the active site has been demonstrated by X-ray crystallographic analysis of *E. coli* dihydrofolate reductase liganded with different cofactors and substrates. In fact, Sawaya and Kraut have analyzed crystal structures of different forms of this protein, including the holo-enzyme, the Michaelis complex, the ternary product complex, the tetrahydrofolate binary complex as well as the tetrahydrofolate-NADPH complex. These structures can be used to reconstruct a 2.1 Å resolution movie, depicting the sequence of events during the catalytic cycle, which showed that the enzyme adopts different conformational states while complexed with different ligands, suggesting that the process of enzymatic catalysis might be accompanied by significant conformational changes [177].

It has been emphasized that signaling and regulation is among the most important function of IDPs [94]. Qualitatively, it seems reasonable that highly mobile proteins would provide a better basis for signaling and recognition. For example, disordered regions can bind partners with both high specificity and low affinity [178]. This means that the regulatory interactions can be specific and also can be easily dispersed. Obviously this represents a keystone of signaling—turning a signal off is as important as turning it on [10]. Another crucial property of IDPs for their function in signaling networks is binding diversity; i.e., their ability to partner with many other proteins and other ligands, such as nucleic acids [179]. This opens a unique possibility for one regulatory region or one regulatory protein to bind to many different partners. In agreement with this hypothesis it has been shown that proteins making multiple interactions are more likely to lead to lethality if deleted [180]. There are several other reasons of why IDPs might be superior to their ordered counterparts. This includes, but is not limited to: binding commonality in which multiple, distinct sequences recognize a common binding site (with perhaps different folds in the various complexed IDPs) [181]; the ability to form large interaction surfaces as the disordered region wraps-up [182] or surrounds its partner [183]; faster rates of association by reducing dependence on orientation factors and by enlarging target sizes [184]; and faster rates of dissociation by unzippering mechanisms [10].

An interesting consequence of the capability of IDPs to interact with different binding partners is their polymorphism in bound state; i.e., an IDP (or ID region) might have completely different geometries in the rigidified structures induced by the binding to its partner, depending on the nature of the bound partner. Recent crystallographic studies on glycogen synthase kinase 3β (GSK3β), a Ser/Thr protein kinase and its interactions with FRAT1 and axin provide an illustrative example of these polymorphic bound states [185]. It has been shown that a sharp turn breaks the structure of FRAT peptide into two distinct and separate α-helical segments, whereas the axin peptide is bound as a single unbroken α-helix with a turn and an irregular tail [185]. Overall, despite the fact that the primary binding sites for axin and FRAT on GSK3β have been found to overlap substantially in the crystal structures, so that their binding is mutually exclusive, the GSK3β-interacting regions of these two proteins were shown to possess negligible sequence similarity [185]. Furthermore, even although both bound peptides are primarily helical, their detailed structures and interactions with GSK3β have substantial differences. It has been emphasized that the ability of GSK3β to bind two different proteins with high specificity via the same binding site is mediated by the conformational plasticity of the 285–299 loop. In fact, in the non-bound GSK3β, this loop is highly mobile and poorly defined. However, it is induced to accommodate one of the two completely distinct and well-ordered structures that are specific to the individual ligand bound: while some residues in this versatile binding site in GSK3β are involved in interactions with both axin and FRAT, others are involved uniquely with one or the other [185].

Based on these and other examples it is clear that being disordered has broad functional importance [8, 10, 12, 51, 52, 92, 94–98, 178, 184, 186]. Many IDPs and IDPRs undergo a disorder-to-order transition upon functioning [8, 10, 12, 51, 52, 92, 94–98, 178, 184, 186, 187]. When disordered regions bind to signaling partners, the free energy required to bring about the disorder to order transition takes away from the interfacial, contact free energy, with the net result that a highly specific interaction can be combined with a low net free energy of association [10, 178]. High specificity coupled with low affinity seems to be a useful pair of properties for a signaling interaction so that the signaling interaction is reversible. In addition, a disordered protein can readily bind to multiple partners by changing shape to associate with different targets [10, 179, 188]. In addition to decoupled specificity and strength of binding, disorder has several clear advantages for functions in signaling, regulation and control [7, 8, 10, 33, 73, 106, 118]:

(1) Increased speed of interaction due to greater capture radius and the ability to spatially search through interaction space;
(2) Strengthened encounter complex allows for less stringent spatial orientation requirements;
(3) Efficient regulation via rapid degradation;
(4) Increased interaction (surface) area per residue;
(5) The ability to be involved in one-to-many binding, where a single disordered region binds to several structurally diverse partners;
(6) The ability to be involved in many-to-one binding, where many distinct (structured) proteins may bind a single disordered region;
(7) The ability to overcome steric restrictions, enabling larger interaction surfaces in protein-protein and protein-ligand complexes than those obtained with rigid partners;
(8) Binding plasticity, where an IDPR folds to specific bound conformations (which can be very different) according to the template provided by binding partners;
(9) Efficient regulation via posttranslational modification; i.e., phosphorylation, methylation, ubiquitination, SUMOylation, etc.;
(10) Ease of regulation/redirection and production of otherwise diverse forms by alternative splicing;
(11) The possibility of overlapping binding sites due to extended linear conformation;

(12) Decoupled binding affinity and specificity, where, due to the induced folding, IDP/IDPR can be involved in the formation of specific but weak complexes. This combination of high specificity with low affinity defines the broad utilization of intrinsic disorder in regulatory interactions where turning a signal off is as important as turning it on [10];

(13) The ability of some IDPs/IDPRs to form very stable intertwined complexes;

(14) Diverse evolutionary rates with some IDPs being highly conserved and other IDPs possessing high evolutionary rates. The latter ones can evolve into sophisticated and complex interaction centers (scaffolds) that can be easily tailored to the needs of divergent organisms;

(15) Flexibility that allows masking (or not) of interaction sites or that allows interaction between bound partners;

(16) The ability to be involved in the cascade interactions, where IDP binding to the first partner induces partial folding generating a new binding site suitable for interaction with the second partner, etc.;

(17) Binding fuzziness where different binding mechanisms (e.g., via stabilizing the binding-competent secondary structure elements within the contacting region, or by establishing the long-range electrostatic interactions, or being involved in transient physical contacts with the partner, or even without any apparent ordering) can be employed to accommodate peculiarities of inter-action with various partners.

This clearly suggests that there is a new two-pathway protein structure-function paradigm, with sequence-to-structure-to-function for enzymes and membrane transport proteins, and sequence-to-disordered ensemble-to-function for proteins and protein regions involved in signaling, regulation, and control [3, 10, 44, 92, 95, 96].

7 Binding Promiscuity and Multitude of the Disorder-Based Binding Modes

IDPs/IDPRs can form highly stable complexes, or be involved in signaling inter-actions where they undergo constant "bound-unbound" transitions, thus acting as dynamic and sensitive "on-off" switches. The ability of these proteins to return to the highly flexible conformations after the completion of a particular function, and their predisposition to gain different conformations depending on the environmental peculiarities, are unique physiological properties of IDPs which allow them to exert different functions in different cellular contests according to a specific conforma-tional state [4].

Static complexes Due to their lack of rigid structure, combined with the high level of intrinsic dynamics and almost unrestricted flexibility at various structure levels in the non-bound state, as well as due to the unique capability to adjust to structure of the binding partner, intrinsically disordered proteins are characterized

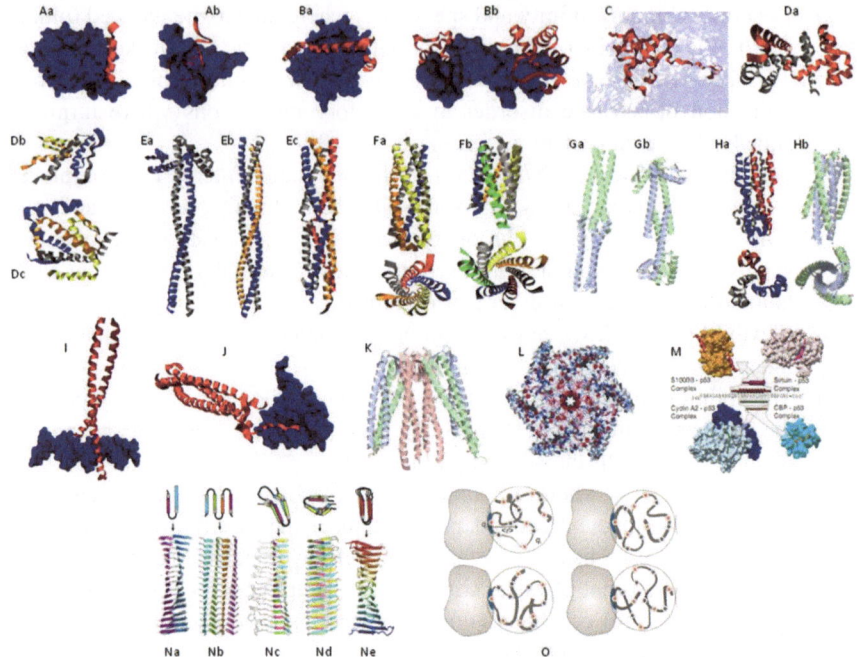

Fig. 9 A portrait gallery of disorder-based complexes. Illustrative examples of various interaction modes of intrinsically disordered proteins are shown. *A* MoRFs. **a** α-MoRF, a complex between the botulinum neurotoxin (*red helix*) and its receptor (a *blue cloud*) (PDB ID: 2NM1); **b** i-MoRF, a complex between an 18-mer cognate peptide derived from the α1 subunit of the nicotinic acetylcholine receptor from *Torpedo californica* (*red helix*) and α-cobratoxin (a *blue cloud*) (PDB ID: 1LXH). *B* Wrappers. **a** rat PP1 (*blue cloud*) complexed with mouse inhibitor-2 (*red helices*) (PDB ID: 2O8A); **b** a complex between the paired domain from the Drosophila paired (prd) protein and DNA (PDB ID: 1PDN). *C* Penetrator. Ribosomal protein s12 embedded into the rRNA (PDB ID: 1N34). *D* Huggers. **a** *E. coli trp* repressor dimer (PDB ID: 1ZT9); **b** tetramerization domain of p53 (PDB ID: 1PES); **c** tetramerization domain of p73 (PDB ID: 2WQI). *E* Intertwined strings. **a** dimeric coiled coil, a basic coiled-coil protein from *Eubacterium eligens* ATCC 27750 (PDB ID: 3HNW); **b** trimeric coiled coil, salmonella trimeric autotransporter adhesin, SadA (PDB ID: 2WPQ); **c** tetrameric coiled coil, the virion-associated protein P3 from Caulimovirus (PDB ID: 2O1J). *F* Long cylindrical containers. **a** pentameric coiled coil, *side* and *top* views of the assembly domain of cartilage oligomeric matrix protein (PDB ID: 1FBM); **b** *side* and *top* views of the seven-helix coiled coil, engineered version of the GCN4 leucine zipper (PDB ID: 2HY6). *G* Connectors. **a** human heat shock factor binding protein 1 (PDB ID: 3CI9); **b** the bacterial cell division protein ZapA from *Pseudomonas aeruginosa* (PDB ID: 1W2E). *H* Armature. **a** *side* and *top* views of the envelope glycoprotein GP2 from Ebola virus (PDB ID: 2EBO); **b**, *side* and *top* views of a complex between the N- and C-terminal peptides derived from the membrane fusion protein of the Visna (PDB ID: 1JEK). *I* Tweezers or forceps. A complex between c-Jun, c-Fos and DNA. Proteins are shown as *red helices*, whereas DNA is shown as a *blue cloud* (PDB ID: 1FOS). *J* Grabbers. Structure of the complex between βPIX coiled coil (*red helices*) and Shank PDZ (*blue cloud*) (PDB ID: 3L4F). *K* Tentacles. Structure of the hexameric molecular chaperone prefoldin from the archaeum *Methanobacterium thermoautotrophicum* (PDB ID: 1FXK). *L* Pullers. Structure of the ClpB chaperone from *Thermus thermophilus* (PDB ID: 1QVR). *M* Chameleons. The C-terminal fragment of p53 gains different types of secondary structure in complexes with four different binding partners, cyclinA (PDB ID: 1H26), sirtuin (PDB ID: 1MA3), CBP bromo domain (PDB

◀ ID: 1JSP), and s100ββ (PDB ID: 1DT7). *N* Stackers or β-arcs. **a** stack of β-arches, β-amyloid; **b** superpleated β-structure (Sup35p, Ure2P, α-synuclein); **c** stack of β-solenoids (prion); **d** stack of β-arch dimers (insulin); **e** β-solenoids. Modified from Ref. [216]. *O* Dynamic complexes. Schematic representation of the polyelectrostatic model of Sic1-Cdc4 interaction. Schematic representation of an IDP (*ribbon*) interacting with a folded receptor (*gray shape*) through several distinct binding motifs and an ensemble of conformations (indicated by four representations of the interaction). The intrinsically disordered protein possesses positive and negative charges (depicted as *blue* and *red circles*, respectively) giving rise to a net charge q_l, while the binding site in the receptor (*light blue*) has a charge q_r. The effective distance $<r>$ is between the binding site and the centre of mass of the intrinsically disordered protein. Modified from Ref. [219] This figure is adapted with permission from [315]. Copyright (2014) American Chemical Society

by a very diverse range of binding modes, creating a multitude of unusual complexes, many of which are not attainable by ordered proteins [189]. Some of these complexes are relatively static, resemble complexes of ordered proteins, and, therefore are suitable for the structure determination by X-ray crystallography. Among these static complexes are: molecular recognition features (MoRFs), wrappers, chameleons, penetrators, huggers, intertwined strings, long cylindrical containers, connectors, armature, tweezers and forceps, grabbers, tentacles, pullers, and stackers or β-arcs [189]. These binding modes are shown in Fig. 9 and briefly described below.

MoRFs MoRFs are short, interaction-prone intrinsically disordered protein segments that undergo disorder-to-order transitions upon binding and are abundantly involved in molecular recognition [190–192]. MoRFs, being identified as short structured fragments of disordered proteins involved in interaction with globular partners, were structurally classified according to their structures in the bound state: α-MoRFs form α-helices, β-MoRFs form β-strands, and ι-MoRFs form structures without a regular pattern of backbone hydrogen bonds [190, 192]. MoRF typically constitutes one contiguous segment fitted into a grove at the surface of the ordered partner.

Flexible Wrappers Some IDPs wrap around their ordered binding partners. Complexes of this type are polyvalent ordered complexes where several ordered segments of a disordered protein bind to disjoint and spatially distant binding sites on the surface of the globular protein. Typically, the ordered segments of such wrapping IDPs are connected by the disordered regions. Secondary structure elements of wrappers almost do not possess intramolecular interactions, forming very intensive intermolecular contacts with binding partner [48, 193–195]. Many proteins interacting with DNA or RNA are flexible wrappers. For example, numerous transcription factors, regulatory proteins, and other proteins that interact with DNA contain multiple zinc finger motifs. The zinc finger motifs act as independently folded globular domains that are separated by flexible linker regions. Zinc finger domains are disordered in the absence of zinc. Proteins often contain multiple zinc fingers connected by flexible linkers and wrap around the DNA in a spiral manner. The zinc finger-containing proteins typically interact with the major groove along the double helix of DNA. In the bound state, the zinc fingers are arranged around the DNA strand in such a way that the α-helix of each finger contacts the DNA, forming an almost continuous stretch of α-helices around the DNA molecule [196].

Penetrators In complexes of some IDPs with other proteins or RNA, significant parts of IDPs penetrate deep inside the structure of their binding partners. For example, in the crystal structure of the 30S ribosome, many ribosomal proteins, in addition to globular domains contained extended internal loops or long N- or C-terminal extensions that were not seen in structures of the isolated proteins, but which were associated intimately with the RNA inside the ribosome [197]. The most illustrative example of this penetrating mode is S12, which had a globular domain at the interface side and a long N-terminal extension that threaded its way through the 30 S subunit to emerge on the back side to interact with proteins S8 and S17 [197].

Huggers Typically, monomers constituting the oligomers formed as a result of folding coupled to binding are highly intertwined [198–200], kind of hugging each other. Typically, complexes of this kind are binary. There are however several examples of group huggers, where monomers clasp more than one partner.

Intertwined strings Coiled coils represent a common structural motif in proteins, where up to 7 long α-helices intertwined together similar to strings of a rope [131, 201]. This motif is formed by approximately 3–5 % of all amino acids in proteins [202], with the most common members of this family being dimers and trimers. Individual α-helices in coiled coil are wrapped around each other into a left-handed helix to form a supercoil. In addition to the left-handed coiled coils there are right-handed coiled coils [203, 204]. Coiled coils represent relatively simple but tightly packed structure. Importantly, partners involved in the coiled-coil formation are typically disordered in the non-bound form.

Long cylindrical containers Multichain coiled coils can assemble into long hollow cylinders containing a continuous axial pore with binding capacities for several hydrophobic compounds [205].

Connectors and armature Being formed, coiled-coils can be used for the subsequent formation of higher order oligomers, where segment of coiled-coils are used to bring oligomeric partners together [206–208]. Coiled coil can serve as an armature, around which more complex structure is built [209, 210].

Tweezers and forceps Many transcription factors form coiled-coil dimers that interact with DNA. Here, the coiled-coil dimer grips the major grove of DNA in a forceps-like manner [211].

Grabbers In several instances, the ends of the coiled-coil form an extensive β-sheet interaction with binding partners [212].

Tentacles In its crystal structure, the hexameric molecular chaperone prefoldin resembles a jellyfish with body consisting of a double β-barrel assembly, from which six long tentacle-like coiled-coils are protruding. The distal regions of the coiled-coils contain hydrophobic patches which are utilized for the multivalent binding of nonnative proteins [213].

Pullers The *Thermus thermophilus* chaperone ClpB is a two-tiered hexameric ring with a set of 85 Å-long and mobile coiled-coils that are located on the outside of the hexamer and act as mechanical pullers [214]. Here, the concerted motions of these coiled-coils causes adjacent subunits to move in opposite directions generating the mechanical force required to pull aggregate apart [214].

Chameleons One of the most unique features of IDPs is their ability to gain, in a template-dependent manner, very different structures in the bond form. This capability is illustrated by the C-terminal binding region of p53, the same short segment of which binds to four unrelated partners adopting different conformations (an α-helix, a β-sheet, and two differently laid irregular structures) when bound to the different partners [4, 110, 215].

Stackers or β-arcs The unifying model of the amyloid fibrils is "β-arcades," which are the columnar structures produced by in-register stacking of "β-arcs", strand-turn-strand motifs in which the two β-strands interact *via* their side chains, not *via* the polypeptide backbone as in a conventional β-hairpin [216].

Disordered or fuzzy complexes of IDPs In addition to the static complexes considered above, where bound partners have fixed structures, some intrinsically disordered proteins do not fold even in their bound state, forming so-called disordered, dynamic, or fuzzy complexes with ordered proteins [217–222], other disordered proteins [223–225], or biological membranes [226, 227]. In complexes of some of these intrinsically disordered proteins with their binding partners, the disordered regions flanking the interaction interface but not the interface itself remain disordered. Such mode of interaction was recently described as "the flanking fuzziness" in contrast to "the random fuzziness" when the disordered protein remains entirely disordered in the bound state [228, 229]. It is also expected that the similar binding mode can be utilized by disordered protein while interacting with nucleic acids and other biological macromolecules [189].

Physically, binding is considered as joining objects together and suggests spatial and temporal fixation of bound partners. The formation of protein complexes with specific binding partners is expected to bring some fixation (at least at the binding site). Therefore, mentioned disordered complexes where interaction of a disordered protein with the binding partners is not accompanied by a disorder-to-order transition within the interaction interface clearly cannot be described by the classical binding paradigm. The contradiction can be resolved assuming that the ordered binding partner and/or disordered protein contain multiple low affinity binding sites. The existence of several similar binding sites combined with a highly flexible and dynamic structure of disordered protein creates a unique situation where any binding site of disordered protein can interact with any binding site of its partner with almost equal probability, in a staccato manner. The low affinity of each individual contact implies that each of them is not stable and can be readily broken. Therefore, such disordered or fuzzy complex can be envisioned as a highly dynamic ensemble in which a disordered protein does not present a single binding site to its partner but resemble a "binding cloud", in which multiple identical binding sites are dynamically distributed in a diffuse manner. In other words, in this staccato-type interaction mode, an disordered protein rapidly changes multiple binding sites while probing binding site(s) of its partner [189]. Additional factor which can help holding dynamic complex together could be a weak long-range attraction between protein molecules [230]. This long-range attraction is universal for all protein solutions and has a range several times that of the diameter of the protein molecule, much greater than the range of the screened electrostatic repulsion [230].

8 IDPs/IDPRs in Human Diseases

Because of the fact that IDP play crucial roles in numerous biological processes, it was not too surprising to find that some of them are involved in human diseases. For example, a number of human diseases originate from the deposition of stable, ordered, filamentous protein aggregates, commonly referred to as amyloid fibrils. In each of these pathological states, a specific protein or protein fragment changes from its natural soluble form into insoluble fibrils, which accumulate in a variety of organs and tissues [231–237]. Approximately 20 different proteins are known so far to be involved in these diseases. These proteins are unrelated in terms of sequence or starting structure. Several IDPs are found in this list of 20, being associated with the development of several neurodegenerative diseases [99, 237]. An incomplete list of disordered associated with IDPs includes Alzheimer's disease (deposition of amyloid-β, tau-protein, α-synuclein fragment NAC [238–241]; Niemann-Pick disease type C, subacute sclerosing panencephalitis, argyrophilic grain disease, myotonic dystrophy, and motor neuron disease with neurofibrillary tangles (accumulation of tau-protein in form of neurofibrillary tangles [240]); Down's syndrome (nonfilamentous amyloid-β deposits [242]); Parkinson's disease, dementia with Lewy body, diffuse Lewy body disease, Lewy body variant of Alzheimer's disease, multiple system atrophy and Hallervorden-Spatz disease (deposition of α-synuclein in a form of Lewy body, or Lewy neuritis [243]); prion diseases (deposition of PrP^{SC}, Ref. [244]); and a family of polyQ diseases, a group of neurodegenerative disorders caused by expansion of GAC trinucleotide repeats coding for PolyQ in the gene products [245]. Furthermore, most mutations in rigid globular proteins associated with accelerated fibrillation and protein deposition diseases have been shown to destabilize the native structure, increasing the steady-state concentration of partially folded (disordered) conformers [231–237].

The disorders given above have been called conformational diseases, as they are characterized by the conformational changes, misfolding and aggregation of an underlying protein. However, there is another side to this coin: protein functionality. In fact, many of the proteins associated with the conformational disorders are also involved in recognition, regulation and cell signaling. For example, functions ascribed to α-synuclein, a protein involved in several neurodegenerative disorders, include binding fatty acids and metal ions; regulation of certain enzymes, transporters, and neurotransmitter vesicles; and regulation of neuronal survival (reviewed in Ref. [243]). Overall, there are more than 50 proteins and ligands that interact and/or co-localize with this protein. Furthermore, α-synuclein has amazing structural plasticity and adopts a series of different monomeric, oligomeric and insoluble conformations (reviewed in Refs. [25, 81, 246, 247]). The choice between these conformations is determined by the peculiarities of the protein environment, assuming that α-synuclein has an exceptional ability to fold in a template-dependent manner. Based on these observations, we hypothesize that the development of the conformational diseases may originate from the misidentification, misregulation and missignaling, accompanied by misfolding. In other words, mutations and/or changes in the environment may result in protein confusion, for which its ID becomes lost, thus reducing its capability to recognize proper binding partners and leading to the formation of nonfunctional and deadly aggregates.

Recent analysis of so-called polyglutamine diseases gives support to this hypothesis [248]. Polyglutamine diseases are a specific group of hereditary neu-rodegeneration caused by expansion of CAG triplet repeats in an exon of disease genes which leads to the production of a disease protein containing an expanded polyglutamine, polyQ, stretch. Nine neurodegenerative disorders, including Ken-nedy's disease, Huntington's diseases, spinocerebellar atrophy-1, -2, -3, -6, 7, 17, and dentatorubral pallidoluysian atrophy are known to belong to this class of dis-eases [249–252]. In most polyQ diseases, expansion to over 40 repeats leads to the onset [252]. It has been emphasized that such molecular processes as unfolded protein response, protein transport, synaptic transmission and transcription are implicated in the pathology of polyQ diseases [248]. Importantly, more than 20 transcription-related factors have been reported to interact with pathological polyQ proteins. Furthermore, these interactions were shown to repress the transcription, leading finally to the neuronal dysfunction and death (reviewed in Ref. [248]). These results suggest that polyQ diseases represent kind of transcriptional disorder [248], supporting our misidentification hypothesis for at least some of the confor-mational disorders.

Disorder is very common in cancer-associated proteins too. In experimental studies, the presence of disorder has been directly observed in several cancer-associated proteins, including p53 [253], $p57^{kip2}$ [254], Bcl-X_L and Bcl-2 [255], c-Fos [256], and a thyroid cancer associated protein, TC-1 [257].

To estimate the abundance of IDPs/IDPRs in various pathological conditions, at least three computational/bioinformatics approaches were elaborated. The first approach is based on the assembly of specific datasets of proteins associated with a given disease and the computational analysis of these datasets using a number of disorder predictors [53, 94, 258–261]. In essence, this is an analysis of individual proteins extended to a set of independent proteins. A second approach utilized diseasome, a network of genetic diseases where the related proteins are interlinked within one disease and between different diseases [262]. A third approach is based on the evaluation of the association between a particular protein function (including the disease-specific functional keywords) with the level of intrinsic disorder in a set of proteins known to carry out this function [100–102]. These three approaches are briefly described below, whereas the results of their application are presented in the subsequent section.

For the first time, the dataset analysis approach was used in 2002 when it was found that 79 % of cancer-associated and 66 % of cell-signaling proteins contain predicted regions of disorder of at least 30 residues [94]. In contrast, only 13 % of a set of proteins with well-defined ordered structures contained such long regions of predicted disorder. For this study, cancer-associated proteins were defined as those human proteins in Swiss-Prot containing the keyword "oncogene" (this included anti- and proto-oncogenes) or containing the word "tumor" in the description field. Following a similar analytical model, a dataset of 487 proteins related to cardiovascular disease (CVD) was collected and analyzed [259]. On average, CVD-related proteins were found to be highly disordered. The percentage of proteins with 30 or more consecutive disordered residues was 61 % for CVD-associated proteins. Many proteins were predicted to be wholly disordered, with 101 proteins from the CVD dataset predicted to have a total of almost 200 specific disorder-based binding motifs (thus about 2 binding sites per protein), α-MoRFs [259]. Finally, the dataset analysis revealed that in addition to being abundant in cancer- and CVD-related proteins, intrinsic disorder is commonly found in such maladies as neurodegenerative diseases and diabetes [260, 263].

The human diseasome systematically links the human disease phenome (which includes 1,284 human genetic diseases, 867 of which had at least one link to other diseases, and 516 diseases formed a giant component) with the human disease genome (which contains 1,777 disease genes of which 1,377 were shown to be connected to other disease genes, and 903 genes belonged to a giant cluster) [264]. The abundance of intrinsic disorder in human diseasome was evaluated using a set of computational tools such as PONDR® VSL2, CDF-analysis, CH-plot, and α-MoRF prediction [262]. These analyses uncovered an unfoldome associated with human genetic diseases and revealed that intrinsic disorder is common in proteins associated with many human genetic diseases. It was also shown that different disease classes vary in the IDP/IDPR contents and that α-MoRFs are common in the diseasome, and their abundance in a given protein correlates with the intrinsic disorder level of this protein. Finally, some disease classes were shown to have a significant fraction of genes affected by alternative splicing, and the alternatively spliced regions in the corresponding proteins are predicted to be highly disordered and in some diseases contain a significant number of MoRFs [262].

The studies on correlation of intrinsic disorder with various functional keywords [100–102] revealed that many diseases were strongly correlated with proteins predicted to be disordered. Contrary to this, no disease-associated proteins were found to be strongly correlated with absence of disorder [102]. Among disease-related Swiss-Prot keywords strongly associated with intrinsic disorder were oncoproteins, malaria, trypanosomiasis, human immunodeficiency virus (HIV) and acquired immunodeficiency syndrome (AIDS), deafness, obesity, cardiovascular disease, diabetes mellitus, albinism, and prion [102]. In agreement with this bioinformatics analysis, at least one illustrative, experimentally validated example of functional disorder or order was found for the vast majority of functional keywords related to diseases [102].

Summarizing, intrinsic disorder is highly abundant among proteins associated with various human diseases. Since IDPs are very common in various diseases, the "disorder in disorders" or D^2 concept was introduced to summarize work in this area [260] and concepts of the disease-related unfoldome and unfoldomics were developed [263].

9 IDPs as Potential Drug Targets

Since many proteins associated with various human diseases either completely disordered or contain long disordered regions [260, 263] and since some of this disease-related intrinsically disordered proteins/regions are involved in recognition, regulation and signaling, these proteins/regions clearly represent novel potential drug targets [28]. It is recognized now that the possibility of interrupting the action of disease-associated proteins (including through modulation of protein-protein interactions) presents an extremely attractive objective for the development of new drugs. The rational design of enzyme inhibitors depends on the classical view of protein function, which states that three-dimensional structure is an obligatory prerequisite for function. While generally applicable to many enzymatic domains, this view has persisted to influence thinking concerning all protein functions despite numerous contrasting examples. Due to failure to recognize the important role of disorder in protein function, current and evolving methods of drug discovery suffer from an overly rigid view of protein function. This is striking in the observation that the vast majority of currently available drugs target the active site of enzymes, presumably since these are the only proteins for which the order-function paradigm is generally applicable.

Disordered proteins often bind their partners via a relatively short length of contiguous residues, which become ordered upon binding [191, 265, 266]. Targeting small molecules to disordered regions of proteins should enable the development of more effective drug discovery techniques. Drugs targeting such regions would likely function through inducing the disordered region to form an ordered structure unlike its structure in complex with its binding partner, thereby preventing association. The principles of small molecule binding to disordered regions have not been well studied, but sequence specific, small molecule binding to short peptides has been observed [267]. An interesting twist of disorder-based approach for drug discovery is that targeting disordered regions as drug targets can be described as inducing order to *prevent* function (or drug-induced misfolding).

In agreement with mentioned above concepts, small molecules, 'Nutlins', have been recently discovered that inhibit the p53-Mdm2 interaction by mimicking the helix in p53 that binds to Mdm2 [268, 269]. The tumor suppressor protein p53 is at the center of a large signaling network. It regulates expression of genes involved in numerous cellular processes, including cell cycle progression, apoptosis induction, DNA repair, as well as others involved in responding to cellular stress [270]. When p53 function is lost, either directly through mutation or indirectly through several other mechanisms, the cell often undergoes cancerous transformation [271, 272]. Cancers showing mutations in p53 are found in colon, lung, esophagus, breast, liver, brain, reticuloendothelial tissues and hemopoietic tissues [271].

p53 is regulated by several different mechanisms including inhibition of its activity by binding to E3 ubiquitin ligase Mdm2, which binds to a short stretch of p53, residues 13–29. This region of p53 is within the transactivation domain, thus p53 cannot activate or inhibit other genes when Mdm2 is bound. Mdm2 ubiquitinates p53 and thus targets it for destruction. Mdm2 also contains a nuclear export signal that causes p53 to be transported out of the nucleus. Although X-ray crystallographic studies of the p53-Mdm2 complex reveal that the Mdm2 binding region of p53 forms a helical structure that binds into a deep groove on the surface of Mdm2 [273], NMR studies show that the unbound N-terminal region of p53 lacks fixed structure, although it does possess an amphipathic helix that forms secondary structure part of the time [253] and is therefore represents an illustrative example of the α-MoRF concept. This amphipathic helix seen in the unbound state is the same helix that binds to Mdm2. A close examination of the interface between the proteins reveals that Phe_{19}, Trp_{23}, and Leu_{26} of p53 are the major contributors to the interaction, with the side chains of these three amino acids pointing down into a crevice on the Mdm2 surface.

Because of the apparent simplicity of the interface, as well as the importance of the p53–Mdm2 interaction, this protein-protein interaction has been investigated as a possible drug target by many researchers. Several successful peptide inhibitors of the interaction have been created [274–277]. These peptides were all derived from the region of p53 that binds to Mdm2. Additionally all successful peptide inhibitors contained the three crucial residues involved in the interaction, Phe_{19}, Trp_{23}, and Leu_{26} [268, 269].

Several small molecules were recently found to block the p53-Mdm2 interaction [269, 278–280]. While some of these were natural products, others were from a class of cis-imidazolines called "Nutlins". These latter molecules increased the level of p53 in cancer cell lines. This drastically decreased the viability of these cells, causing most of them to undergo apoptosis. When one of the Nutlin compounds was given orally to mice, researchers saw a 90 % inhibition of tumor growth compared to the control. The structure of Nutlin-2 was shown to mimic the crucial residues of p53, with two bromophenyl groups fitting into Mdm2 in the same pockets as Trp_{23} and Leu_{26}, and an ethyl-ether side chain filling the spot normally taken by Phe_{19} [278–280].

This research demonstrates that finding small molecules to target regions of proteins normally bound by disordered proteins is an important future challenge. It is anticipated that by studying this drug interaction, researchers will be able to identify regions of other proteins that can be mimicked by small molecules. Remarkably, the disorder prediction for p53 using PONDR® VL-XT software showed a sharp downward spike indicating predicted ordered region near the N-terminus of the protein. Furthermore, the α-MoRF identifier was able to recognize the region of p53 that binds to Mdm2 as a region of molecular recognition [110].

This successful nutlin story marks the potential beginning of a new era, *the signaling-modulation era*, in targeting drugs to protein-protein interactions. Importantly, this druggable p53-Mdm2 interaction involves a disorder-to-order transition. Principles of such a transition are generally understood and therefore can use to find similar drug targets [281]. In addition to nutlins, seven types of promising drug molecules that act by blocking protein-protein interactions have been described [282, 283]. While protein disorder is not mentioned in any of the papers describing how a small molecule can block protein-protein interactions, the

disorder-based analysis revealed that four of these interactions involve one structured partner and one disordered partner, with 3 of the 4 disordered segments
becoming helix upon binding. Therefore, the p53—Mdm2 complex is not the only
member of this class currently known to be blocked by a small drug-like molecule.
We fully expect many more examples to appear shortly, and for some of these
examples to lead to useful drug molecules. Since p53-mdm2-like interactions are
likely to be very common, they clearly define a cornucopia of new drug targets that
would operate by blocking disorder-based protein-protein interactions.

For these examples, the drug molecules mimic a critical region of the disordered
partner (which folds upon binding). Such drugs can efficiently compete with this
foldable disordered region for its binding site on the structured partner. We argue
that these druggable sites are likely to operate by a coupled binding and folding
mechanism and utilize interaction sites that are small enough and compact enough
to be easily mimicked by small molecules [28]. We have developed methods for
predicting such binding sites in disordered regions [284] and have elaborated on the
bioinformatics tools to identify which disordered binding regions can be easily
mimicked by small molecules [281].

A complementary approach for finding small molecules inhibiting disorder-based
protein interactions has been developed in the lab of Prof. Steven Metallo (see e.g.
Ref. [285]). Deregulation of the c-Myc transcription factor is involved in many types
of cancer, making this oncoprotein an attractive target for drug discovery. In order to
bind DNA, regulated target gene expression, and function in most biological con-
texts, c-Myc must dimerize with its obligate heterodimerization partner, Max, which
lacks a transactivation segment. c-Myc, which is intrinsically disordered as a
monomer, undergoes coupled binding and folding of its basic-helix-loop-helix-
leucine zipper domain (bHLHZip) upon heterodimerization with its partner protein
Max. One approach to c-Myc inhibition has been to disrupt this dimeric complex. In
a search for effective inhibitors of the c-Myc-Max interactions, a series of small
molecules were analyzed to find seven Myc inhibitors, which were shown to bind to
one of three discrete sites within the 85-residue bHLHZip domain of c-Myc. These
binding sites are composed of short contiguous stretches of amino acids that can
selectively and independently bind small molecules. Inhibitor binding induces only
local conformational changes, preserves the overall disorder of c-Myc, and inhibits
dimerization with Max. Furthermore, binding of inhibitors to c-Myc was shown to
occur simultaneously and independently on the three independent sites. Based on
these observations it has been concluded that a rational and generic approach to the
inhibition of protein-protein interactions involving IDPs may therefore be possible
through the targeting of intrinsically disordered sequence [285].

Recently, a functional misfolding concept was introduced to describe a mechanism preventing IDPs from unwanted interactions with non-native partners [145].
IDPs/IDPRs are characterized by high conformational dynamics and flexibility, the
presence of sticky preformed binding elements, and the ability to morph into dif-
ferently-shaped bound configurations. However, detailed analyses of the conformational behavior and fine structure of several IDPs revealed that the preformed
binding elements might be involved in a set of non-native intramolecular

interactions. Based on these observations it was proposed that an intrinsically disordered polypeptide chain in its unbound state can be misfolded to sequester the preformed elements inside the non-interactive or less-interactive cage, therefore preventing these elements from the unnecessary and unwanted interactions with non-native binding partners [145]. It is important to remember, however, that the mentioned functional misfolding is related to the ensemble behavior of transiently populated elements of structure. In other words, it describes the behavior of a globally disordered polypeptide chain containing highly dynamic elements of residual structure, so-called interaction-prone preformed fragments, some of which could potentially be related to protein function [145].

This ability of IDRPs/IDPRs to functionally misfold (i.e., to spontaneously form a non-interactive cage sequestering interaction-prone preformed fragments) can be exploited in the drug discovery process for finding small molecules which would potentially stabilize different members of the functionally misfolded ensemble, and therefore prevents the targeted protein from establishing biological interactions [115]. This approach is very different from the discussed above direct targeting of short IDPRs since it is based on a small molecule binding to a highly dynamic surface created via the transient interaction of preformed interaction-prone fragments. In essence, this approach can be considered as an extension of the well-established structure-based rational drug design elaborated for ordered proteins. In fact, if the structure of a member(s) of the functionally misfolded ensemble can be guessed, then this structure can be used to find small molecules that are potentially able to interact with this structure, utilizing tools originally developed for the rational structure-based drug design for ordered proteins [286].

Ideally, a drug that targets a given protein-protein interaction should be tissue specific. Although some proteins are unique for a given tissue, many more proteins have very wide distribution, being present in several tissues and organs. How can one develop tissue-specific drugs targeting such abundant proteins? Often, tissue specificity for many of the abundant proteins is achieved via the alternative splicing of the corresponding pre-mRNAs, which generates two or more protein isoforms from a single gene. Estimates indicate that between 35 and 60 % of human genes yield protein isoforms by means of alternatively spliced mRNA [149]. The added protein diversity from alternative splicing is thought to be important for tissue-specific signaling and regulatory networks in the multicellular organisms. Recently, it has been established that the regions of alternative splicing in proteins are enriched in intrinsic disorder [54]. Since disorder is frequently utilized in protein binding regions, having alternative splicing of pre-mRNA coupled to regions of protein disorder was proposed to lead to tissue-specific signaling and regulatory diversity [54]. Therefore, associating alternative splicing with protein disorder enables the time- and tissue-specific modulation of protein function. These findings open a unique opportunity to develop tissue-specific drugs modulating the function of a given IDP/IDPR (with a unique profile of disorder distribution) in a target tissue and not affecting the functionality of this same protein (with different disorder distribution profile) in other tissues.

10 Concluding Remarks: Intrinsic Disorder as a Universal Tool for Solving Protein Mysteries and Riddles

Several examples considered in these Briefs represent perfect illustrations of how the use of the protein intrinsic disorder concept might help in solving protein-related mysteries. Let us consider several of these cases.

Intrinsic disorder and PTMs The first example is posttranslational modifications (PTMs), where there is an obvious conflicts between the lock-and-key mechanism of enzymatic activity and high promiscuity of modifying enzymes (which often have multiple targets), and between the evolutionary-shaped folding mechanism supporting the "unique sequence-unique structure" concept and the fact that PTMs dramatically change physico-chemical properties of the modified residues. In fact, it was pointed out that in any given proteome, the number of kinases and phosphatases is noticeably smaller than the number of their potential substrates, and each eukaryotic protein kinase serves ~ 20 substrates, whereas each human phosphatase is expected to dephosphorylate ~ 65 clients. The reality is further complicated due to the abundant presence of multiple phosphorylation sites in one protein, which are often targeted by several different kinases and phosphatases. In other words, the classical lock-and-key mode of the enzyme action is converted here to the one-lock-many-keys scenario from the kinase/phosphatase side or to the one key-many-locks scenario from the side of the target protein. To further complicate this picture, the activity of many kinases serve is controlled and regulated via autophosphorylation and/or phosphorylation by other kinases. Similar situation is expected for other enzymes responsible for various posttranslational modifications of proteins.

Since PTMs can occur at any stage of protein's life, and since PTMs are responsible for dramatic extension of the physico-chemical properties of residues (due to the variety of PTMs, the available arsenal of residues increases from 20 primary residues encoded by DNA to ~ 140), another important side of this story is the effect of PTMs on structure and folding of a target protein. It is believed that the ability of a given polypeptide chain to fold into specific structure is encoded in its specific sequence of amino acids which was evolutionary selected to provide optimal folding. In other words, according to the classic structure-function paradigm, due to the evolutionary selection, any given polypeptide chain "knows" how to fold into unique, biologically active structure. Furthermore, the protein structure and the mechanism of protein folding can be dramatically altered by single point mutations, especially by those substitutions which would result in large changes in the physico-chemical properties of the substituted residue (e.g., changing hydrophobic residues to charged residues, etc.). Although PTM is not a genetic mutation, this process will undoubtedly change chemical structure and physico-chemical properties of the modified residue. One can argue that PTM typically occurs posttranslationally, when protein is already folded, and therefore, the corresponding modification of a residue should not have a direct effect on protein folding. However, even under the physiologic conditions, ordered proteins are known to exist in the dynamic equilibrium between folded and partially folded conformations due to the conformational breathing determined by the fact that the conformational forces stabilizing the protein structure are weak and can be broken even at ambient temperatures due to the thermal fluctuations. Therefore, some PTMs are expected to affect this equilibrium, potentially possessing dramatic effects on the outputs of the folding process leading to some serious structural damages.

Obviously, both of these obstacles (the inability of classic "lock-and-key" model to explain the mechanisms of action of enzymes responsible for PTMs and the potential detrimental effects of PTMs on folding and structure of the modified proteins) can be easily avoided if the sites targeted for modifications are located within the IDPRs. In fact, flexibility of a region surrounding modification site of a target protein will allow this region to serve as a flexible lock pick able to adjust itself to the peculiarities of the active site. Also, the lack of stable structure in such a region eliminates the PTM/protein structure/folding conundrum since an entity without structure cannot be broken. As it follows from various studies, coupling of PTM sites with disorder in the surrounding region is what Nature does not only to avoid mentioned problems but also to ease the access of the modifying enzymes to the sites targeted for modification.

Intrinsic disorder and alternative splicing Alternative splicing (AS) leading to the extension of the functional proteome during gene expression by providing means to generate several proteins from one gene represents another challenge to the classic structure-function paradigm. Although general logistics of this paradigm seems to be applicable to this important mechanism (to extend functional repertoire of a given gene, multiple mRNA sequences of various length are generated which encode proteins of various length), AS clearly might have detrimental outputs on protein folding. As it was already mentioned, the efficient folding of ordered proteins is a result of evolutionary selection to find polypeptide chains capable of independent folding to a specific functional structure. Even if AS would not affect the reading frame, this process could induce folding disaster. This hypothesis is based on the fact that AS ultimately generates polypeptide chains of different length where different structural segments are removed. The question now is the resistance of the protein structure to such interventions. Consequences of the deletion of a part of structure are rather obvious: (a) protein structure will survive if a taken out part is a short fragment of just a few residues, or a part of a surface loop, or an independent domain; (b) protein folding would be challenged in all other cases as well in cases when AS results in the reading frame shift. With exclusive prevalence of AS in high eukaryotes where almost all genes are subjected to alternative splicing, the chance of generating "folding disaster" leading to misfolded and potentially pathogenic species is very high. This seems to be a very dangerous way of proteome extension where for each functional protein form there is a chance to generate several misfolded entities. As it was pointed out (see above, *Functional regulation via alternative splicing* section), this riddle is solved by linking AS with intrinsic disorder, where mRNA regions affected by AS encode for IDPRs. Again, no folding catastrophe can be expected for a structure-less protein/region.

Intrinsic disorder in hub and scaffold proteins Hub and scaffold proteins are involved in one-to-many interactions, which are difficult to explain using the

(a)   **(b)**

Fig. 10 How would rigid hubs and their rigid partners look like if the classic "lock-and-key" model would be utilized? **a** A structured one-to-many binder suitable for interaction with a set of structured partners would look like a key-ring with multiple keys. **b** An ordered many-to-one binder that is be able to interact with a set of ordered binding partners would look like as multiple locks chained together

"lock-and-key" model. In fact, the more realistic description of the one-to-many interactions between rigid (or structured) partners would be scenarios involving loss of interaction specificity (e.g., one lock accepting many keys, or one master key opening many locks), or scenarios with preserved specificity but highly increased complexity (e.g., a key-ring with multiple keys, or multiple locks chained together (see Fig. 10)). Since neither of these scenarios represents an optimal solution of this problem, Nature is using intrinsic disorder in hub/scaffold proteins and/or their partners to ensure efficient one-to-many binding.

Intrinsic disorder and entropic chain activities Some IDP/IDPR functions do not involve interactions and utilize the highly flexible and entropic nature of these extended protrusions. Among these entropy chain functions ascribed to IDTPs are entropic bristle and entropic clock activities [117].

In its role of an entropic bristle, an IDPR would sweep out a significant region in space by random movements about its point of attachment and entropically exclude large particles without excluding small molecules such as water, salts, metals, or cofactors, therefore acting as entropic bristle or entropic bristle domain [287]. Illustrative examples of natural entropic bristles are disordered termini of neuro-filament H and M proteins that are >300 and >600 residues long, respectively, and that occupy space by thermally driven motion and thereby maintain the separation of neighboring filaments, and also maintains the axonal bore, possibly allowing the movement of small molecules and maintaining the shape of the axon against compression [288]. Recently, the ability of highly disordered tails to serve as entropic bristles (EBs) preventing proteins from interaction was used to design of the EB-based protein solubilizers that extend away from the partner and sweep out large molecules, therefore allowing the target protein to fold free from interference [289]. Here, a set of natural and artificial EBs (that have a low level of sequence complexity, a high net charge and are diversified by means of distinctive amino acid compositions and lengths) was designed as intrinsically disordered, highly charged protein sequences that were translationally fused to partner proteins to serve as effective solubilizers by creating both a large favorable surface area for water interactions and large excluded volumes around the partner [289].

IDPTs are also crucial components of the entropic clocks that provide a timing mechanism arising from random searches such as those observed in the ball-and-chain model for closure of voltage-gated ion channels. In the Shaker potassium channel, a specific region located at the intrinsically disordered N-terminal tail was shown to play an important role in inactivation by stochastically interacting with the open channel to cause inactivation [124]. Recently, a detailed analysis revealed that there is a linear correlation between the length of the channel's tail and its binding affinity to the scaffold protein partner PSD-95, whereas the dissociation rate constant is independent on chain length [290]. These observations suggested that the IRPT of the Shaker channel controls the entropy of association whereas the PDZ binding motif located at the tip of the tail controls the interaction enthalpy. This provided a strong support to the entropic clock mechanism, where the disordered tail of the channel modulates the timing of the complex formation with the scaffold protein partner [290]. Obviously, the efficiency of the entropic chain activities is directly related to intrinsic disorder, and ordered proteins/domains cannot serve as entropic bristles or entropic clocks.

Intrinsic disorder and "funny" proteins. In addition to the very general cases considered above where intrinsic disorder provided solution to some global problems of some global biological processes, the use of intrinsic disorder concept represents an attractive way to solve mysteries and riddles of individual proteins too. Examples of this are vast, and only two cases are given below as characteristic illustrations.

It is clear now that the amino acid sequence peculiarities of IDPs/IDPRs can be blamed for their unusual and unexpected behavior. Many early IDP researchers were stunned by the peculiar features of these mysterious then members of the protein kingdom. On a personal note [91], my journey to the IDP field started when one sunny day, an excited colleague of mine appeared in the lab shaking a tube with a sample in his hand and shouting: "I have a funny protein here. I cannot measure its concentration. And it is extremely stable—I can boil it for a few days, but as soon as I am bringing temperature down it shows 100 % activity". That funny protein was prothymosin α. Even superficial analysis of the amino acid sequence of this protein clearly showed that the unusual behavior of this protein is definitely determined by its peculiar amino acid composition. It does not have any aromatic residues and cysteins. Therefore its concentration cannot be measured spectroscopically. 64 of 111 residues in this protein have charged side groups (there are 19 Asp, 35 Glu, 2 Arg, and 8 Lys residues), whereas overall content of hydrophobic residues (Leu, Ile and Val) is very low [291]. Based on this amino acid composition, it was not a big surprise to find that prothymosin α behaved as a highly disordered coil-like chain—you cannot expect that highly charged polypeptide (60 % polyE/D) will have a strong tendency to fold under the physiological conditions [91]. This lack of stable structure also explained extreme thermal and acid stability of prothymosin α—you cannot break what is already broken [291].

Another illustration is actin, which was recently proposed to be a hybrid protein containing ordered and disordered regions [292, 293]. Actin is one of the most abundant proteins in the eukaryotic cells, where the monomeric actin concentrations are within the 12–300 μM range [294]. Being found in almost every living cell, this globular multifunctional protein is most common in muscle cells, where its concentration ranges from 230 to 960 μM [294]. Among various functional and structural features ascribed to actin are its ability to bind one divalent cation and one molecule of ATP (or ADP), the ability to exists as a monomer known as G-actin (globular form which is populated under low ionic strength conditions) or a single-stranded polymer, the so called fibrous form of actin, or F-actin (fibrillar form, which results from the polymerization of G-actin in the presence of neutral salts), or an inactive form (I-actin) that lacks the ability to polymerize [295–297].

Assignment of actin to the category of IDPs is not a trivial statement since this globular protein is known to possess 3D-structure. However, the available structural information about this protein is derived from its complexes with various actin binding proteins (ABPs), such as DNase I (PDB ID: 1ATN) [298], a *Vibrio parahaemolyticus* effector protein VopL (PDB ID: 4M63) [299], chimera of gelsolin domain 1 and C-terminal domain of thymosin β-4 (PDB ID: 1T44) [300], and many other proteins. One of the potential reasons for inability of actin to be crystallized by its own can be attributed to the strong intrinsic propensity of G-actin to polymerize into the F-actin. Although this high polymerization tendency precludes G-actin from the crystal formation, it can be prevented via binding of this protein to some ABPs. As a result, actin can be crystallized in the presence of these ABPs. In addition to the G- and F-forms, actin can be easily transformed into I-actin, biologically inactive form unable to polymerize. The transition to the I-actin can be initiated by the removal of calcium ion by the EDTA or EGTA treatment, removal of nucleotides (ATP or ADP), heat denaturation, exposure to moderate urea or GdnHCl concentrations, dialysis with 8 M urea or 6 M GdnHCl, or spontaneously during storage [295, 301–307]. This inactivated actin is characterized by the intrinsic fluorescence spectrum with maximum at wavelength intermediate between the wavelengths of the native and completely unfolded protein

[304], combined with rather rigid microenvironment of tryptophan residues [308], a considerable increase of the fluorescence anisotropy value reflecting a considerable decrease in the internal mobility of the tryptophan residues in the inactivated actin [307], and a noticeable distortion of the secondary structure [308].

Furthermore, actin possesses a very unusual unfolding behavior, which can be described as unfolding with a trap [309, 310]. Here, the existence of the clear minima in the kinetic profiles describing unfolding of this protein in the range of 1.0–2.0 M GdnHCl suggested that the transition from the native to the inactivated state occurs via some essentially unfolded intermediate state of actin, U*, which is a mostly unfolded kinetic intermediate, the fluorescent properties of which are similar to those of the completely unfolded state but which possesses noticeable amount of ordered secondary structure [309, 310]. Later stopped-flow-based analysis revealed the existence of several additional intermediate states [311, 312]. These studies gave raise to the unusual kinetic model of actin unfolding [310]:

$$
\begin{array}{c}
 \overset{[\text{GdnHCl}] < 1.8\text{M}}{} \overset{[\text{GdnHCl}]=0.2\text{M}}{} \\
N \longrightarrow N^* \longrightarrow U^* \rightleftharpoons I_1 \ldots \rightleftharpoons I_n \ldots \rightleftharpoons I \rightleftharpoons I_{\text{aggr.}} \\
\Big\updownarrow \; {\scriptstyle [\text{GdnHCl}] > 3.0\text{M}} I=I_{15} \text{ at } [\text{GdnHCl}]=0\text{M.} \\
U
\end{array}
$$

According to this model, the transition state N^* precedes the transformation of native actin into the essentially unfolded state (U^*). The formation of this essentially unfolded state (U^*) precedes the formation of completely unfolded (U) or inactivated actin (I). In the processes of folding and unfolding, the essentially unfolded state (U^*) is an on-pathway intermediate, whereas inactivated actin (I) is an off-pathway associate, the appearance of which competes with the transition to the native state [310]. In this scheme, I_{aggr} corresponds to the aggregated form of inactivated actin.

Based on these and other observations, it is evident that actin meets the majority of the characteristics of IDPs. In fact, like some other IDPs, actin cannot fold into a compact state without chaperones. This protein not only cannot fold without chaperones but also cannot form a compact structure without its ligands, the Ca^{2+} ion and ATP. In vitro, unfolding of this protein is an irreversible process, indicating that the information encoded in its polypeptide chain is not sufficient to ensure normal folding. Also, it cannot maintain the folded native state without fastening it with Ca^{2+} ions. Actin always exists in complexes: while folding, it successively interacts with the chaperone Hsp 70, then with prefoldin (PFD) and finally with the chaperonin CCT, which provides for correct folding and Ca^{2+} and ATP incorporation; fibrillar actin is formed by the self-association of G-actin molecules; in the cytoplasm or nucleus, actin is in complex with ABPs; and, in particular, the G-actin pool is preserved in complex with profilin. Interestingly, inactivated actin is also a monodisperse oligomer (not an amorphous aggregate) that, possibly, has some functional role.

As many other IDPs, actin interacts with an enormous number of partners [313] and possesses numerous PTM sites. While interacting with various ABPs, actin acts as a hub protein, as is typical for IDPs. Many of the ABPs themselves are IDPs and are involved in various signaling system and are known to interact with other hub proteins. Actin is ubiquitous and multifunctional protein. It is one of the main components of the system of muscle contraction, it forms the cytoskeleton, it is found in the cell nucleus in which, except for the motility and scaffold functions, actin acts as a regulator protein that participates in the processes of transcription and chromatin remodeling.

References

1. Fischer, E. *Ber. Dt. Chem. Ges.* **1894**, *27*, 2985.
2. Bernstein, F. C.; Koetzle, T. F.; Williams, G. J.; Meyer, E. F., Jr.; Brice, M. D.; Rodgers, J. R.; Kennard, O.; Shimanouchi, T.; Tasumi, M. *J Mol Biol* **1977**, *112*, 535.
3. Uversky, V. N. *Protein Sci* **2002**, *11*, 739.
4. Uversky, V. N.; Dunker, A. K. *Biochim Biophys Acta* **2010**, *1804*, 1231.
5. Uversky, V. N. *Protein Sci* **2013**, *22*, 693.
6. Petsko, G. A.; Ringe, D. *Primers in Biology. Protein Structure and Function.*; New Science Press Ltd., Sinauer Associates, Inc. Publishers, Blackwell Publishing: London, 2004.
7. Dunker, A. K.; Garner, E.; Guilliot, S.; Romero, P.; Albrecht, K.; Hart, J.; Obradovic, Z.; Kissinger, C.; Villafranca, J. E. *Pac Symp Biocomput* **1998**, 473.
8. Wright, P. E.; Dyson, H. J. *J Mol Biol* **1999**, *293*, 321.
9. Uversky, V. N.; Gillespie, J. R.; Fink, A. L. *Proteins* **2000**, *41*, 415.
10. Dunker, A. K.; Lawson, J. D.; Brown, C. J.; Williams, R. M.; Romero, P.; Oh, J. S.; Oldfield, C. J.; Campen, A. M.; Ratliff, C. M.; Hipps, K. W.; Ausio, J.; Nissen, M. S.; Reeves, R.; Kang, C.; Kissinger, C. R.; Bailey, R. W.; Griswold, M. D.; Chiu, W.; Garner, E. C.; Obradovic, Z. *J Mol Graph Model* **2001**, *19*, 26.
11. Tompa, P. *Trends Biochem Sci* **2002**, *27*, 527.
12. Daughdrill, G. W.; Pielak, G. J.; Uversky, V. N.; Cortese, M. S.; Dunker, A. K. In *Handbook of Protein Folding*; Buchner, J., Kiefhaber, T., Eds.; Wiley-VCH, Verlag GmbH & Co. KGaA: Weinheim, Germany, 2005.
13. Dunker, A. K.; Babu, M. M.; Barbar, E.; Blackledge, M.; Bondos, S. E.; Dosztányi, Z.; Dyson, H. J.; Forman-Kay, J.; Fuxreiter, M.; Gsponer, J.; Han, K.-H.; Jones, D. T.; Longhi, S.; Metallo, S. J.; Nishikawa, K.; Nussinov, R.; Obradovic, Z.; Pappu, R.; Rost, B.; Selenko, P.; Subramaniam, V.; Sussman, J. L.; Tompa, P.; Uversky, V. N. *Intrinsically Disordered Proteins* **2013**, *1*, e24157.
14. Dunker, A. K.; Obradovic, Z.; Romero, P.; Garner, E. C.; Brown, C. J. *Genome Inform Ser Workshop Genome Inform* **2000**, *11*, 161.
15. Ward, J. J.; Sodhi, J. S.; McGuffin, L. J.; Buxton, B. F.; Jones, D. T. *J Mol Biol* **2004**, *337*, 635.
16. Uversky, V. N. *J Biomed Biotechnol* **2010**, *2010*, 568068.
17. Xue, B.; Dunker, A. K.; Uversky, V. N. *J Biomol Struct Dyn* **2012**, *30*, 137.
18. Holt, C.; Sawyer, L. *J Chem Soc Faraday Trans* **1993**, *89*, 2683.
19. Pullen, R. A.; Jenkins, J. A.; Tickle, I. J.; Wood, S. P.; Blundell, T. L. *Mol. Cell. Biochem.* **1975**, *8*, 5.
20. Cary, P. D.; Moss, T.; Bradbury, E. M. *Eur. J. Biochem.* **1978**, *89*, 475.
21. Linderstrom-Lang, K.; Schellman, J. A. In *The Enzymes*; 2nd ed.; Boyer, P. D., Lardy, H., Myrback, K., Eds.; Academic Press: New York, 1959.
22. Schweers, O.; Schonbrunn-Hanebeck, E.; Marx, A.; Mandelkow, E. *J Biol Chem* **1994**, *269*, 24290.

V.N. Uversky, *Intrinsically Disordered Proteins*, Protein Folding and Structure,
DOI 10.1007/978-3-319-08921-8

23. Weinreb, P. H.; Zhen, W.; Poon, A. W.; Conway, K. A.; Lansbury, P. T., Jr. *Biochemistry* **1996**, *35*, 13709.

24. Chen, J.; Liang, H.; Fernandez, A. *Genome Biol* **2008**, *9*, R107.

25. Uversky, V. N. *J Biomol Struct Dyn* **2003**, *21*, 211.

26. Fuxreiter, M.; Tompa, P.; Simon, I.; Uversky, V. N.; Hansen, J. C.; Asturias, F. J. *Nat Chem Biol* **2008**, *4*, 728.

27. Tsvetkov, P.; Asher, G.; Paz, A.; Reuven, N.; Sussman, J. L.; Silman, I.; Shaul, Y. *Proteins* **2008**, *70*, 1357.

28. Dunker, A. K.; Uversky, V. N. *Curr Opin Pharmacol* **2010**, *10*, 782.

29. Livesay, D. R. *Curr Opin Pharmacol* **2010**, *10*, 706.

30. Janin, J.; Sternberg, M. J. E. *F1000 Biol Rep* **2013**, *5*, 2.

31. Campen, A.; Williams, R. M.; Brown, C. J.; Meng, J.; Uversky, V. N.; Dunker, A. K. *Protein Pept Lett* **2008**, *15*, 956.

32. Radivojac, P.; Iakoucheva, L. M.; Oldfield, C. J.; Obradovic, Z.; Uversky, V. N.; Dunker, A. K. *Biophys J* **2007**, *92*, 1439.

33. Romero, P.; Obradovic, Z.; Li, X.; Garner, E. C.; Brown, C. J.; Dunker, A. K. *Proteins* **2001**, *42*, 38.

34. Garner, E.; Cannon, P.; Romero, P.; Obradovic, Z.; Dunker, A. K. *Genome Inform Ser Workshop Genome Inform* **1998**, *9*, 201.

35. Williams, R. M.; Obradovi, Z.; Mathura, V.; Braun, W.; Garner, E. C.; Young, J.; Takayama, S.; Brown, C. J.; Dunker, A. K. *Pac Symp Biocomput* **2001**, 89.

36. Vacic, V.; Uversky, V. N.; Dunker, A. K.; Lonardi, S. *BMC Bioinformatics* **2007**, *8*, 211.

37. Li, X.; Obradovic, Z.; Brown, C. J.; Garner, E. C.; Dunker, A. K. *Genome Inform Ser Workshop Genome Inform* **2000**, *11*, 172.

38. Ferron, F.; Longhi, S.; Canard, B.; Karlin, D. *Proteins* **2006**, *65*, 1.

39. Bourhis, J. M.; Canard, B.; Longhi, S. *Curr Protein Pept Sci* **2007**, *8*, 135.

40. Dosztanyi, Z.; Sandor, M.; Tompa, P.; Simon, I. *Curr Protein Pept Sci* **2007**, *8*, 161.

41. Dosztanyi, Z.; Tompa, P. *Methods Mol Biol* **2008**, *426*, 103.

42. He, B.; Wang, K.; Liu, Y.; Xue, B.; Uversky, V. N.; Dunker, A. K. *Cell Res* **2009**, *19*, 929.

43. Jin, F.; Liu, Z. *Biophys J* **2013**, *104*, 488.

44. Uversky, V. N. *Eur J Biochem* **2002**, *269*, 2.

45. Romero, P.; Obradovic, Z.; Kissinger, C. R.; Villafranca, J. E.; Garner, E.; Guilliot, S.; Dunker, A. K. *Pac Symp Biocomput* **1998**, 437.

46. Feng, Z. P.; Zhang, X.; Han, P.; Arora, N.; Anders, R. F.; Norton, R. S. *Mol Biochem Parasitol* **2006**, *150*, 256.

47. Tompa, P.; Dosztanyi, Z.; Simon, I. *J Proteome Res* **2006**, *5*, 1996.

48. Galea, C. A.; High, A. A.; Obenauer, J. C.; Mishra, A.; Park, C. G.; Punta, M.; Schlessinger, A.; Ma, J.; Rost, B.; Slaughter, C. A.; Kriwacki, R. W. *Journal of proteome research* **2009**, *8*, 211.

49. Xue, B.; Williams, R. W.; Oldfield, C. J.; Dunker, A. K.; Uversky, V. N. *BMC Syst Biol* **2010**, *4 Suppl 1*, S1.

50. Burra, P. V.; Kalmar, L.; Tompa, P. *PLoS One* **2010**, *5*, e12069.

51. Dunker, A. K.; Cortese, M. S.; Romero, P.; Iakoucheva, L. M.; Uversky, V. N. *FEBS J* **2005**, *272*, 5129.

52. Uversky, V. N.; Oldfield, C. J.; Dunker, A. K. *J Mol Recognit* **2005**, *18*, 343.

53. Mohan, A.; Sullivan, W. J., Jr.; Radivojac, P.; Dunker, A. K.; Uversky, V. N. *Mol Biosyst* **2008**, *4*, 328.

54. Romero, P. R.; Zaidi, S.; Fang, Y. Y.; Uversky, V. N.; Radivojac, P.; Oldfield, C. J.; Cortese, M. S.; Sickmeier, M.; LeGall, T.; Obradovic, Z.; Dunker, A. K. *Proc Natl Acad Sci U S A* **2006**, *103*, 8390.

55. Oparin, A. I. *The Origin of Life (in Russian)*; Moscow Worker publisher: Moscow, 1924

56. Haldane, J. B. S. In *The Rationalist Annual for the Year 1929*; Watts, C. A., Ed.; Watts & Co: London 1929.

57. Miller, S. L. *Science* **1953**, *117*, 528.
58. Miller, S. L.; Urey, H. C. *Science* **1959**, *130*, 245.
59. Crick, F. H. *J Mol Biol* **1968**, *38*, 367.
60. Wong, J. T. *Proceedings of the National Academy of Sciences of the United States of America* **1975**, *72*, 1909.
61. Jukes, T. H. *Nature* **1973**, *246*, 22.
62. Trifonov, E. N. *Gene* **2000**, *261*, 139.
63. Poole, A. M.; Jeffares, D. C.; Penny, D. *J Mol Evol* **1998**, *46*, 1.
64. Jeffares, D. C.; Poole, A. M.; Penny, D. *J Mol Evol* **1998**, *46*, 18.
65. Tompa, P.; Csermely, P. *Faseb J* **2004**, *18*, 1169.
66. Treiber, D. K.; Williamson, J. R. *Curr Opin Struct Biol* **2001**, *11*, 309.
67. Cristofari, G.; Darlix, J. L. *Prog Nucleic Acid Res Mol Biol* **2002**, *72*, 223.
68. Gilbert, W. *Nature* **1986**, *319*, 618.
69. Csermely, P. *Trends in Biochemical Sciences* **1997**, *22*, 147.
70. Doolittle, W. F. *Sci Am* **2000**, *282*, 90.
71. Theobald, D. L. *Nature* **2010**, *465*, 219.
72. Chemes, L. B.; Glavina, J.; Alonso, L. G.; Marino-Buslje, C.; de Prat-Gay, G.; Sanchez, I. E. *PLoS One* **2012**, *7*, e47661.
73. Brown, C. J.; Takayama, S.; Campen, A. M.; Vise, P.; Marshall, T. W.; Oldfield, C. J.; Williams, C. J.; Dunker, A. K. *J Mol Evol* **2002**, *55*, 104.
74. Chen, S. C.; Chuang, T. J.; Li, W. H. *Mol Biol Evol* **2011**, *28*, 2513.
75. Lin, Y. S.; Hsu, W. L.; Hwang, J. K.; Li, W. H. *Mol Biol Evol* **2007**, *24*, 1005.
76. Shaiu, W. L.; Hu, T.; Hsieh, T. S. *Pac Symp Biocomput* **1999**, 578.
77. Sayers, E. W.; Gerstner, R. B.; Draper, D. E.; Torchia, D. A. *Biochemistry* **2000**, *39*, 13602.
78. Wissmann, R.; Baukrowitz, T.; Kalbacher, H.; Kalbitzer, H. R.; Ruppersberg, J. P.; Pongs, O.; Antz, C.; Fakler, B. *J Biol Chem* **1999**, *274*, 35521.
79. Brown, C. J.; Johnson, A. K.; Dunker, A. K.; Daughdrill, G. W. *Curr Opin Struct Biol* **2011**, *21*, 441.
80. Nilsson, J.; Grahn, M.; Wright, A. P. *Genome Biol* **2011**, *12*, R65.
81. Breydo, L.; Wu, J. W.; Uversky, V. N. *Biochim Biophys Acta* **2012**, *1822*, 261.
82. Uversky, V. N. *J Neurochem* **2007**, *103*, 17.
83. George, J. M. *Genome Biol* **2002**, *3*, REVIEWS3002.
84. Vonderviszt, F.; Kanto, S.; Aizawa, S.; Namba, K. *J Mol Biol* **1989**, *209*, 127.
85. Chen, J. W.; Romero, P.; Uversky, V. N.; Dunker, A. K. *J Proteome Res* **2006**, *5*, 879.
86. Chen, J. W.; Romero, P.; Uversky, V. N.; Dunker, A. K. *J Proteome Res* **2006**, *5*, 888.
87. Balazs, A.; Csizmok, V.; Buday, L.; Rakacs, M.; Kiss, R.; Bokor, M.; Udupa, R.; Tompa, K.; Tompa, P. *FEBS J* **2009**, *276*, 3744.
88. Xue, B.; Brown, C. J.; Dunker, A. K.; Uversky, V. N. *Biochim Biophys Acta* **2013**, *1834*, 725.
89. Daughdrill, G. W.; Narayanaswami, P.; Gilmore, S. H.; Belczyk, A.; Brown, C. J. *J Mol Evol* **2007**, *65*, 277.
90. Moesa, H. A.; Wakabayashi, S.; Nakai, K.; Patil, A. *Mol Biosyst* **2012**, *8*, 3262.
91. Uversky, V. N. *Biochim Biophys Acta* **2013**.
92. Dunker, A. K.; Obradovic, Z. *Nat Biotechnol* **2001**, *19*, 805.
93. Ma, B.; Kumar, S.; Tsai, C. J.; Nussinov, R. *Protein Eng* **1999**, *12*, 713.
94. Iakoucheva, L. M.; Brown, C. J.; Lawson, J. D.; Obradovic, Z.; Dunker, A. K. *J Mol Biol* **2002**, *323*, 573.
95. Dunker, A. K.; Brown, C. J.; Lawson, J. D.; Iakoucheva, L. M.; Obradovic, Z. *Biochemistry* **2002**, *41*, 6573.
96. Dunker, A. K.; Brown, C. J.; Obradovic, Z. *Adv Protein Chem* **2002**, *62*, 25.
97. Dyson, H. J.; Wright, P. E. *Curr Opin Struct Biol* **2002**, *12*, 54.
98. Dyson, H. J.; Wright, P. E. *Nat Rev Mol Cell Biol* **2005**, *6*, 197.

99. Uversky, V. N.; Fink, A. L. In *Amyloid Proteins: The Beta Pleated Sheet Conformation and Disease*; Sipe, J. D., Ed.; Wiley-VCH, Verlag GmbH & Co. KGaA: Weinheim, Germany, 2005.

100. Xie, H.; Vucetic, S.; Iakoucheva, L. M.; Oldfield, C. J.; Dunker, A. K.; Uversky, V. N.; Obradovic, Z. *J Proteome Res* **2007**, *6*, 1882.

101. Vucetic, S.; Xie, H.; Iakoucheva, L. M.; Oldfield, C. J.; Dunker, A. K.; Obradovic, Z.; Uversky, V. N. *J Proteome Res* **2007**, *6*, 1899.

102. Xie, H.; Vucetic, S.; Iakoucheva, L. M.; Oldfield, C. J.; Dunker, A. K.; Obradovic, Z.; Uversky, V. N. *J Proteome Res* **2007**, *6*, 1917.

103. Uversky, V. N. *Cell Mol Life Sci* **2003**, *60*, 1852.

104. Tompa, P. *FEBS Lett* **2005**, *579*, 3346.

105. Tompa, P.; Szasz, C.; Buday, L. *Trends Biochem Sci* **2005**, *30*, 484.

106. Cortese, M. S.; Uversky, V. N.; Dunker, A. K. *Prog Biophys Mol Biol* **2008**, *98*, 85.

107. Dunker, A. K.; Silman, I.; Uversky, V. N.; Sussman, J. L. *Curr Opin Struct Biol* **2008**, *18*, 756.

108. Dunker, A. K.; Oldfield, C. J.; Meng, J.; Romero, P.; Yang, J. Y.; Chen, J. W.; Vacic, V.; Obradovic, Z.; Uversky, V. N. *BMC Genomics* **2008**, *9 Suppl 2*, S1.

109. Dunker, A. K.; Uversky, V. N. *Nat Chem Biol* **2008**, *4*, 229.

110. Oldfield, C. J.; Meng, J.; Yang, J. Y.; Yang, M. Q.; Uversky, V. N.; Dunker, A. K. *BMC Genomics* **2008**, *9 Suppl 1*, S1.

111. Russell, R. B.; Gibson, T. J. *FEBS Lett* **2008**, *582*, 1271.

112. Tompa, P.; Fuxreiter, M.; Oldfield, C. J.; Simon, I.; Dunker, A. K.; Uversky, V. N. *Bioessays* **2009**, *31*, 328.

113. Uversky, V. N.; Dunker, A. K. *Science* **2008**, *322*, 1340.

114. Wright, P. E.; Dyson, H. J. *Curr Opin Struct Biol* **2009**, *19*, 31.

115. Uversky, V. N. *Int J Biochem Cell Biol* **2012**, *43*, 1090.

116. Uversky, V. N. *Curr Pharm Des* **2013**, *42*, 4191–4213.

117. Uversky, V. N. *FEBS Lett* **2013**, *587*, 1891.

118. Dunker, A. K.; Obradovic, Z.; Romero, P.; Kissinger, C.; Villafranca, E. *PDB Newsletter* **1997**, *81*, 3.

119. Antz, C.; Geyer, M.; Fakler, B.; Schott, M. K.; Guy, H. R.; Frank, R.; Ruppersberg, J. P.; Kalbitzer, H. R. *Nature* **1997**, *385*, 272.

120. Armstrong, C. M.; Bezanilla, F. *J Gen Physiol* **1977**, *70*, 567.

121. Herson, P. S.; Virk, M.; Rustay, N. R.; Bond, C. T.; Crabbe, J. C.; Adelman, J. P.; Maylie, J. *Nat Neurosci* **2003**, *6*, 378.

122. Lerche, H.; Jurkat-Rott, K.; Lehmann-Horn, F. *Am J Med Genet* **2001**, *106*, 146.

123. Liebovitch, L. S.; Selector, L. Y.; Kline, R. P. *Biophys J* **1992**, *63*, 1579.

124. Hoshi, T.; Zagotta, W. N.; Aldrich, R. W. *Science* **1990**, *250*, 533.

125. Erdös, P.; Rényi, A. *Publ. Math. Inst. Hung. Acad. Sci.* **1960**, *5*, 17.

126. Barabasi, A. L.; Bonabeau, E. *Sci Am* **2003**, *288*, 60.

127. Watts, D. J.; Strogatz, S. H. *Nature* **1998**, *393*, 440.

128. Goh, K. I.; Oh, E.; Jeong, H.; Kahng, B.; Kim, D. *Proceedings of the National Academy of Sciences of the United States of America* **2002**, *99*, 12583.

129. Barabasi, A. L.; Albert, R. *Science* **1999**, *286*, 509.

130. Ekman, D.; Light, S.; Bjorklund, A. K.; Elofsson, A. *Genome Biol* **2006**, *7*, R45.

131. Liu, W.; Rui, H.; Wang, J.; Lin, S.; He, Y.; Chen, M.; Li, Q.; Ye, Z.; Zhang, S.; Chan, S. C.; Chen, Y. G.; Han, J.; Lin, S. C. *Embo J* **2006**, *25*, 1646.

132. Liu, J.; Perumal, N. B.; Oldfield, C. J.; Su, E. W.; Uversky, V. N.; Dunker, A. K. *Biochemistry* **2006**, *45*, 6873.

133. Minezaki, Y.; Homma, K.; Kinjo, A. R.; Nishikawa, K. *J Mol Biol* **2006**, *359*, 1137.

134. Toth-Petroczy, A.; Oldfield, C. J.; Simon, I.; Takagi, Y.; Dunker, A. K.; Uversky, V. N.; Fuxreiter, M. *PLoS Comput Biol* **2008**, *4*, e1000243.

135. Peng, Z.; Mizianty, M. J.; Xue, B.; Kurgan, L.; Uversky, V. N. *Mol Biosyst* **2012**, *8*, 1886.

136. Peng, Z.; Xue, B.; Mizianty, M. J.; Oldfield, C. J.; Kurgan, L.; Uversky, V. N. *Biochim. Biophys. Acta - Molecular Cell Research* **2013**, In press.
137. Westerheide, S. D.; Raynes, R.; Powell, C.; Xue, B.; Uversky, V. N. *Curr Protein Pept Sci* **2012**, *13*, 86.
138. Xue, B.; Oldfield, C. J.; Van, Y. Y.; Dunker, A. K.; Uversky, V. N. *Mol Biosyst* **2012**, *8*, 134.
139. Xue, B.; Dunker, A. K.; Uversky, V. N. *J Biomol Struct Dyn* **2012**, *29*, 843.
140. Peng, Z.; Xue, B.; Kurgan, L.; Uversky, V. N. *Cell Death & Differentiation* **2013**, In press.
141. Papadakos, G.; Housden, N. G.; Lilly, K. J.; Kaminska, R.; Kleanthous, C. *J Mol Biol* **2012**, *418*, 269.
142. Bonsor, D. A.; Grishkovskaya, I.; Dodson, E. J.; Kleanthous, C. *J Am Chem Soc* **2007**, *129*, 4800.
143. Bonsor, D. A.; Hecht, O.; Vankemmelbeke, M.; Sharma, A.; Krachler, A. M.; Housden, N. G.; Lilly, K. J.; James, R.; Moore, G. R.; Kleanthous, C. *EMBO J* **2009**, *28*, 2846.
144. Collins, E. S.; Whittaker, S. B.; Tozawa, K.; MacDonald, C.; Boetzel, R.; Penfold, C. N.; Reilly, A.; Clayden, N. J.; Osborne, M. J.; Hemmings, A. M.; Kleanthous, C.; James, R.; Moore, G. R. *J Mol Biol* **2002**, *318*, 787.
145. Uversky, V. N. *Biochimica Et Biophysica Acta-Proteins and Proteomics* **2011**, *1814*, 693.
146. Sambrook, J. *Nature* **1977**, *268*, 101.
147. Black, D. L. *Annu Rev Biochem* **2003**, *72*, 291.
148. Ast, G. *Nat Rev Genet* **2004**, *5*, 773.
149. Stamm, S.; Ben-Ari, S.; Rafalska, I.; Tang, Y.; Zhang, Z.; Toiber, D.; Thanaraj, T. A.; Soreq, H. *Gene* **2005**, *344*, 1.
150. Brett, D.; Hanke, J.; Lehmann, G.; Haase, S.; Delbruck, S.; Krueger, S.; Reich, J.; Bork, P. *FEBS Lett* **2000**, *474*, 83.
151. Johnson, J. M.; Castle, J.; Garrett-Engele, P.; Kan, Z.; Loerch, P. M.; Armour, C. D.; Santos, R.; Schadt, E. E.; Stoughton, R.; Shoemaker, D. D. *Science* **2003**, *302*, 2141.
152. Graveley, B. R. *Trends Genet* **2001**, *17*, 100.
153. Minneman, K. P. *Molecular interventions* **2001**, *1*, 108.
154. Thai, T. H.; Kearney, J. F. *J Immunol* **2004**, *173*, 4009.
155. Scheper, W.; Zwart, R.; Baas, F. *Neurogenetics* **2004**, *5*, 223.
156. Wang, P.; Yan, B.; Guo, J. T.; Hicks, C.; Xu, Y. *Proceedings of the National Academy of Sciences of the United States of America* **2005**, *102*, 18920.
157. Furnham, N.; Ruffle, S.; Southan, C. *Proteins* **2004**, *54*, 596.
158. Deng, C. X.; Brodie, S. G. *Bioessays* **2000**, *22*, 728.
159. Mark, W. Y.; Liao, J. C.; Lu, Y.; Ayed, A.; Laister, R.; Szymczyna, B.; Chakrabartty, A.; Arrowsmith, C. H. *J Mol Biol* **2005**, *345*, 275.
160. Walsh, C. T.; Garneau-Tsodikova, S.; Gatto, G. J., Jr. *Angew Chem Int Ed Engl* **2005**, *44*, 7342.
161. Witze, E. S.; Old, W. M.; Resing, K. A.; Ahn, N. G. *Nat Methods* **2007**, *4*, 798.
162. Baumann, M.; Meri, S. *Expert Rev Proteomics* **2004**, *1*, 207.
163. Deribe, Y. L.; Pawson, T.; Dikic, I. *Nat Struct Mol Biol* **2010**, *17*, 666.
164. Mann, M.; Jensen, O. N. *Nat Biotechnol* **2003**, *21*, 255.
165. Yang, X. J. *Oncogene* **2005**, *24*, 1653.
166. Mersfelder, E. L.; Parthun, M. R. *Nucleic Acids Res* **2006**, *34*, 2653.
167. Marks, F. *Protein Phosphorylation*; VCH Weinheim: New York, Basel, Cambridge, Tokyo, 1996.
168. Bossemeyer, D.; Engh, R. A.; Kinzel, V.; Ponstingl, H.; Huber, R. *EMBO J.* **1993**, *12*, 849.
169. Narayana, N.; Cox, S.; Shaltiel, S.; Taylor, S. S.; Xuong, N. *Biochemistry* **1997**, *36*, 4438.
170. Lowe, E. D.; Noble, M. E.; Skamnaki, V. T.; Oikonomakos, N. G.; Owen, D. J.; Johnson, L. N. *Embo. J.* **1997**, *16*, 6646.
171. ter Haar, E.; Coll, J. T.; Austen, D. A.; Hsiao, H. M.; Swenson, L.; Jain, J. *Nat. Struct. Biol.* **2001**, *8*, 593.
172. Hubbard, S. R. *Embo. J.* **1997**, *16*, 5572.

173. McDonald, I. K.; Thornton, J. M. *J. Mol. Biol.* **1994**, *238*, 777.
174. Iakoucheva, L. M.; Radivojac, P.; Brown, C. J.; O'Connor, T. R.; Sikes, J. G.; Obradovic, Z.; Dunker, A. K. *Nucleic Acids Res.* **2004**, *32*, 1037.
175. Radivojac, P.; Vacic, V.; Haynes, C.; Cocklin, R. R.; Mohan, A.; Heyen, J. W.; Goebl, M. G.; Iakoucheva, L. M. *Proteins* **2010**, *78*, 365.
176. Koshland, D. E., Jr. *Proc Natl Acad Sci USA* **1958**, *44*, 98.
177. Sawaya, M. R.; Kraut, J. *Biochemistry* **1997**, *36*, 586.
178. Schulz, G. E. In *Molecular Mechanism of Biological Recognition*; Balaban, M., Ed.; Elsevier/North-Holland Biomedical Press: New York, 1979.
179. Kriwacki, R. W.; Hengst, L.; Tennant, L.; Reed, S. I.; Wright, P. E. *Proceedings of the National Academy of Sciences of the United States of America* **1996**, *93*, 11504.
180. Jeong, H.; Mason, S. P.; Barabasi, A. L.; Oltvai, Z. N. *Nature* **2001**, *411*, 41.
181. Iakoucheva, L. M.; Kimzey, A. L.; Masselon, C. D.; Bruce, J. E.; Garner, E. C.; Brown, C. J.; Dunker, A. K.; Smith, R. D.; Ackerman, E. J. *Protein Sci* **2001**, *10*, 560.
182. Choo, Y.; Schwabe, J. W. *Nat Struct Biol* **1998**, *5*, 253.
183. Meador, W. E.; Means, A. R.; Quiocho, F. A. *Science* **1992**, *257*, 1251.
184. Pontius, B. W. *Trends Biochem Sci* **1993**, *18*, 181.
185. Dajani, R.; Fraser, E.; Roe, S. M.; Yeo, M.; Good, V. M.; Thompson, V.; Dale, T. C.; Pearl, L. H. *Embo J* **2003**, *22*, 494.
186. Plaxco, K. W.; Gross, M. *Nature* **1997**, *386*, 657.
187. Spolar, R. S.; Record, M. T., Jr. *Science* **1994**, *263*, 777.
188. Karush, F. *J. Am. Chem. Soc.* **1950**, *72*, 2705.
189. Uversky, V. N. *Chem Soc Rev* **2011**.
190. Mohan, A. Master of Science, Indiana University, 2006.
191. Oldfield, C. J.; Cheng, Y.; Cortese, M. S.; Romero, P.; Uversky, V. N.; Dunker, A. K. *Biochemistry* **2005**, *44*, 12454.
192. Vacic, V.; Oldfield, C. J.; Mohan, A.; Radivojac, P.; Cortese, M. S.; Uversky, V. N.; Dunker, A. K. *J Proteome Res* **2007**, *6*, 2351.
193. Galea, C. A.; Nourse, A.; Wang, Y.; Sivakolundu, S. G.; Heller, W. T.; Kriwacki, R. W. *J Mol Biol* **2008**, *376*, 827.
194. Galea, C. A.; Wang, Y.; Sivakolundu, S. G.; Kriwacki, R. W. *Biochemistry* **2008**, *47*, 7598.
195. Graham, T. A.; Weaver, C.; Mao, F.; Kimelman, D.; Xu, W. *Cell* **2000**, *103*, 885.
196. Luscombe, N. M.; Austin, S. E.; Berman, H. M.; Thornton, J. M. *Genome Biol* **2000**, *1*, REVIEWS001.
197. Brodersen, D. E.; Clemons, W. M., Jr.; Carter, A. P.; Wimberly, B. T.; Ramakrishnan, V. *J Mol Biol* **2002**, *316*, 725.
198. Teschke, C. M.; King, J. *Curr Opin Biotechnol* **1992**, *3*, 468.
199. Xu, D.; Tsai, C. J.; Nussinov, R. *Protein Sci* **1998**, *7*, 533.
200. Gunasekaran, K.; Tsai, C. J.; Nussinov, R. *J Mol Biol* **2004**, *341*, 1327.
201. Mason, J. M.; Arndt, K. M. *Chembiochem* **2004**, *5*, 170.
202. Wolf, E.; Kim, P. S.; Berger, B. *Protein Sci* **1997**, *6*, 1179.
203. Harbury, P. B.; Plecs, J. J.; Tidor, B.; Alber, T.; Kim, P. S. *Science* **1998**, *282*, 1462.
204. Stetefeld, J.; Jenny, M.; Schulthess, T.; Landwehr, R.; Engel, J.; Kammerer, R. A. *Nat Struct Biol* **2000**, *7*, 772.
205. Ozbek, S.; Engel, J.; Stetefeld, J. *Embo J* **2002**, *21*, 5960.
206. Low, H. H.; Moncrieffe, M. C.; Lowe, J. *J Mol Biol* **2004**, *341*, 839.
207. Zhao, X.; Ghaffari, S.; Lodish, H.; Malashkevich, V. N.; Kim, P. S. *Nat Struct Biol* **2002**, *9*, 117.
208. Liu, X.; Xu, L.; Liu, Y.; Tong, X.; Zhu, G.; Zhang, X. C.; Li, X.; Rao, Z. *Proteins* **2009**, *75*, 1.
209. Malashkevich, V. N.; Schneider, B. J.; McNally, M. L.; Milhollen, M. A.; Pang, J. X.; Kim, P. S. *Proceedings of the National Academy of Sciences of the United States of America* **1999**, *96*, 2662.

210. Malashkevich, V. N.; Singh, M.; Kim, P. S. *Proceedings of the National Academy of Sciences of the United States of America* **2001**, *98*, 8502.
211. Glover, J. N.; Harrison, S. C. *Nature* **1995**, *373*, 257.
212. Im, Y. J.; Kang, G. B.; Lee, J. H.; Park, K. R.; Song, H. E.; Kim, E.; Song, W. K.; Park, D.; Eom, S. H. *J Mol Biol* **2010**, *397*, 457.
213. Siegert, R.; Leroux, M. R.; Scheufler, C.; Hartl, F. U.; Moarefi, I. *Cell* **2000**, *103*, 621.
214. Lee, S.; Sowa, M. E.; Watanabe, Y. H.; Sigler, P. B.; Chiu, W.; Yoshida, M.; Tsai, F. T. *Cell* **2003**, *115*, 229.
215. Hsu, W. L.; Oldfield, C. J.; Xue, B.; Meng, J.; Huang, F.; Romero, P.; Uversky, V. N.; Dunker, A. K. *Protein Sci* **2013**, *22*, 258.
216. Kajava, A. V.; Baxa, U.; Steven, A. C. *Faseb J* **2010**, *24*, 1311.
217. Borg, M.; Mittag, T.; Pawson, T.; Tyers, M.; Forman-Kay, J. D.; Chan, H. S. *Proceedings of the National Academy of Sciences of the United States of America* **2007**, *104*, 9650.
218. Mittag, T.; Orlicky, S.; Choy, W. Y.; Tang, X.; Lin, H.; Sicheri, F.; Kay, L. E.; Tyers, M.; Forman-Kay, J. D. *Proceedings of the National Academy of Sciences of the United States of America* **2008**, *105*, 17772.
219. Mittag, T.; Kay, L. E.; Forman-Kay, J. D. *J Mol Recognit* **2010**, *23*, 105.
220. Mittag, T.; Marsh, J.; Grishaev, A.; Orlicky, S.; Lin, H.; Sicheri, F.; Tyers, M.; Forman-Kay, J. D. *Structure* **2010**, *18*, 494.
221. Sigalov, A. B.; Kim, W. M.; Saline, M.; Stern, L. J. *Biochemistry* **2008**, *47*, 12942.
222. Permyakov, S. E.; Millett, I. S.; Doniach, S.; Permyakov, E. A.; Uversky, V. N. *Proteins* **2003**, *53*, 855.
223. Sigalov, A.; Aivazian, D.; Stern, L. *Biochemistry* **2004**, *43*, 2049.
224. Sigalov, A. B.; Zhuravleva, A. V.; Orekhov, V. Y. *Biochimie* **2007**, *89*, 419.
225. Pometun, M. S.; Chekmenev, E. Y.; Wittebort, R. J. *J Biol Chem* **2004**, *279*, 7982.
226. Sigalov, A. B.; Hendricks, G. M. *Biochem Biophys Res Commun* **2009**, *389*, 388.
227. Sigalov, A. B.; Aivazian, D. A.; Uversky, V. N.; Stern, L. J. *Biochemistry* **2006**, *45*, 15731.
228. Tompa, P.; Fuxreiter, M. *Trends Biochem Sci* **2008**, *33*, 2.
229. Hazy, E.; Tompa, P. *Chemphyschem* **2009**, *10*, 1415.
230. Liu, Y.; Fratini, E.; Baglioni, P.; Chen, W. R.; Chen, S. H. *Phys Rev Lett* **2005**, *95*, 118102.
231. Kelly, J. W. *Curr Opin Struct Biol* **1998**, *8*, 101.
232. Dobson, C. M. *Trends Biochem Sci* **1999**, *24*, 329.
233. Bellotti, V.; Mangione, P.; Stoppini, M. *Cell Mol Life Sci* **1999**, *55*, 977.
234. Uversky, V. N.; Talapatra, A.; Gillespie, J. R.; Fink, A. L. *Med. Sci. Monitor* **1999**, *5*, 1001.
235. Uversky, V. N.; Talapatra, A.; Gillespie, J. R.; Fink, A. L. *Med. Sci. Monitor* **1999**, *5*, 1238.
236. Rochet, J. C.; Lansbury, P. T., Jr. *Curr Opin Struct Biol* **2000**, *10*, 60.
237. Uversky, V. N.; Fink, A. L. *Biochim Biophys Acta* **2004**, *1698*, 131.
238. Glenner, G. G.; Wong, C. W. *Biochem Biophys Res Commun* **1984**, *122*, 1131.
239. Masters, C. L.; Multhaup, G.; Simms, G.; Pottgiesser, J.; Martins, R. N.; Beyreuther, K. *Embo J* **1985**, *4*, 2757.
240. Lee, V. M.; Balin, B. J.; Otvos, L., Jr.; Trojanowski, J. Q. *Science* **1991**, *251*, 675.
241. Ueda, K.; Fukushima, H.; Masliah, E.; Xia, Y.; Iwai, A.; Yoshimoto, M.; Otero, D. A.; Kondo, J.; Ihara, Y.; Saitoh, T. *Proceedings of the National Academy of Sciences of the United States of America* **1993**, *90*, 11282.
242. Wisniewski, K. E.; Dalton, A. J.; McLachlan, C.; Wen, G. Y.; Wisniewski, H. M. *Neurology* **1985**, *35*, 957.
243. Dev, K. K.; Hofele, K.; Barbieri, S.; Buchman, V. L.; van der Putten, H. *Neuropharmacology* **2003**, *45*, 14.
244. Prusiner, S. B. *N Engl J Med* **2001**, *344*, 1516.
245. Zoghbi, H. Y.; Orr, H. T. *Curr Opin Neurobiol* **1999**, *9*, 566.
246. Silva, B. A.; Breydo, L.; Uversky, V. N. *Mol Neurobiol* **2012**, *47*, 446.
247. Uversky, V. N.; Eliezer, D. *Curr Protein Pept Sci* **2009**, *10*, 483.
248. Okazawa, H. *Cell Mol Life Sci* **2003**, *60*, 1427.

249. Cummings, C. J.; Zoghbi, H. Y. *Hum Mol Genet* **2000**, *9*, 909.

250. Gusella, J. F.; MacDonald, M. E. *Nat Rev Neurosci* **2000**, *1*, 109.

251. Orr, H. T. *Genes Dev* **2001**, *15*, 925.

252. Fischbeck, K. H. *Brain Res Bull* **2001**, *56*, 161.

253. Lee, H.; Mok, K. H.; Muhandiram, R.; Park, K. H.; Suk, J. E.; Kim, D. H.; Chang, J.; Sung, Y. C.; Choi, K. Y.; Han, K. H. *J Biol Chem* **2000**, *275*, 29426.

254. Adkins, J. N.; Lumb, K. J. *Proteins* **2002**, *46*, 1.

255. Chang, B. S.; Minn, A. J.; Muchmore, S. W.; Fesik, S. W.; Thompson, C. B. *Embo J* **1997**, *16*, 968.

256. Campbell, K. M.; Terrell, A. R.; Laybourn, P. J.; Lumb, K. J. *Biochemistry* **2000**, *39*, 2708.

257. Sunde, M.; McGrath, K. C.; Young, L.; Matthews, J. M.; Chua, E. L.; Mackay, J. P.; Death, A. K. *Cancer Res* **2004**, *64*, 2766.

258. Uversky, V. N.; Roman, A.; Oldfield, C. J.; Dunker, A. K. *J Proteome Res* **2006**, *5*, 1829.

259. Cheng, Y.; LeGall, T.; Oldfield, C. J.; Dunker, A. K.; Uversky, V. N. *Biochemistry* **2006**, *45*, 10448.

260. Uversky, V. N.; Oldfield, C. J.; Dunker, A. K. *Annu Rev Biophys* **2008**, *37*, 215.

261. Uversky, V. N. *Front Biosci* **2009**, *14*, 5188.

262. Midic, U.; Oldfield, C. J.; Dunker, A. K.; Obradovic, Z.; Uversky, V. N. *PLoS Computational Biology* **2008**, In press.

263. Uversky, V. N.; Oldfield, C. J.; Midic, U.; Xie, H.; Xue, B.; Vucetic, S.; Iakoucheva, L. M.; Obradovic, Z.; Dunker, A. K. *BMC Genomics* **2009**, *10 Suppl 1*, S7.

264. Goh, K. I.; Cusick, M. E.; Valle, D.; Childs, B.; Vidal, M.; Barabasi, A. L. *Proceedings of the National Academy of Sciences of the United States of America* **2007**, *104*, 8685.

265. Garner, E.; Romero, P.; Dunker, A. K.; Brown, C.; Obradovic, Z. *Genome Inform Ser Workshop Genome Inform* **1999**, *10*, 41.

266. Cheng, Y.; Oldfield, C. J.; Meng, J.; Romero, P.; Uversky, V. N.; Dunker, A. K. *Biochemistry* **2007**, *46*, 13468.

267. Chen, C. T.; Wagner, H.; Still, W. C. *Science* **1998**, *279*, 851.

268. Chene, P. *Cell Cycle* **2004**, *3*, 460.

269. Chene, P. *Mol Cancer Res* **2004**, *2*, 20.

270. Anderson, C. W.; Appella, E. In *Handbook of Cell Signaling*; Bradshaw, R. A., Dennis, E. A., Eds.; Academic Press: New York, 2004.

271. Hollstein, M.; Sidransky, D.; Vogelstein, B.; Harris, C. C. *Science* **1991**, *253*, 49.

272. Balint, E. E.; Vousden, K. H. *Br J Cancer* **2001**, *85*, 1813.

273. Kussie, P. H.; Gorina, S.; Marechal, V.; Elenbaas, B.; Moreau, J.; Levine, A. J.; Pavletich, N. P. *Science* **1996**, *274*, 948.

274. Bottger, A.; Bottger, V.; Sparks, A.; Liu, W. L.; Howard, S. F.; Lane, D. P. *Curr Biol* **1997**, *7*, 860.

275. Wasylyk, C.; Salvi, R.; Argentini, M.; Dureuil, C.; Delumeau, I.; Abecassis, J.; Debussche, L.; Wasylyk, B. *Oncogene* **1999**, *18*, 1921.

276. Chene, P.; Fuchs, J.; Bohn, J.; Garcia-Echeverria, C.; Furet, P.; Fabbro, D. *J Mol Biol* **2000**, *299*, 245.

277. Garcia-Echeverria, C.; Chene, P.; Blommers, M. J.; Furet, P. *J Med Chem* **2000**, *43*, 3205.

278. Klein, C.; Vassilev, L. T. *Br J Cancer* **2004**, *91*, 1415.

279. Vassilev, L. T. *Cell Cycle* **2004**, *3*, 419.

280. Vassilev, L. T.; Vu, B. T.; Graves, B.; Carvajal, D.; Podlaski, F.; Filipovic, Z.; Kong, N.; Kammlott, U.; Lukacs, C.; Klein, C.; Fotouhi, N.; Liu, E. A. *Science* **2004**, *303*, 844.

281. Cheng, Y.; LeGall, T.; Oldfield, C. J.; Mueller, J. P.; Van, Y. Y.; Romero, P.; Cortese, M. S.; Uversky, V. N.; Dunker, A. K. *Trends Biotechnol* **2006**, *24*, 435.

282. Arkin, M. R.; Wells, J. A. *Nat Rev Drug Discov* **2004**, *3*, 301.

283. Arkin, M. *Curr Opin Chem Biol* **2005**, *9*, 317.

284. Cochran, A. G. *Chem Biol* **2000**, *7*, R85.

285. Hammoudeh, D. I.; Follis, A. V.; Prochownik, E. V.; Metallo, S. J. *J Am Chem Soc* **2009**, *131*, 7390.
286. Uversky, V. N. *Expert Opin Drug Discov* **2012**, *7*, 475.
287. Hoh, J. H. *Proteins* **1998**, *32*, 223.
288. Brown, H. G.; Hoh, J. H. *Biochemistry* **1997**, *36*, 15035.
289. Santner, A. A.; Croy, C. H.; Vasanwala, F. H.; Uversky, V. N.; Van, Y. Y.; Dunker, A. K. *Biochemistry* **2012**, *51*, 7250.
290. Zandany, N.; Magidovich, E.; Marciano, S.; Orr, I.; Abdu, U.; Yifrach, O. *Biophysical Journal* **2013**, *104*, 466a.
291. Uversky, V. N.; Gillespie, J. R.; Millett, I. S.; Khodyakova, A. V.; Vasiliev, A. M.; Chernovskaya, T. V.; Vasilenko, R. N.; Kozlovskaya, G. D.; Dolgikh, D. A.; Fink, A. L.; Doniach, S.; Abramov, V. M. *Biochemistry* **1999**, *38*, 15009.
292. Na, I.; Reddy, K. D.; Breydo, L.; Xue, B.; Uversky, V. N. *Molecular BioSystems* **2014**, In press.
293. Povarova, O. I.; Uversky, V. N.; Kuznetsova, I. M.; Turoverov, K. K. *Intrinsically Disordered Proteins* **2014**.
294. Pollard, T. D.; Blanchoin, L.; Mullins, R. D. *Annu. Rev. Biophys. Biomol. Struct.* **2000**, *29*, 545.
295. Lehrer, S. S.; Kerwar, G. *Biochemistry* **1972**, *11*, 1211.
296. Strzelecka-Golaszewska, H.; Nagy, B.; Gergely, J. *Arch. Biochem. Biophys.* **1974**, *161*, 559.
297. Strzelecka-Golaszewska, H.; Venyaminov, S.; Zmorzynski, S.; Mossakowska, M. *Eur. J. Biochem.* **1985**, *147*, 331.
298. Kabsch, W.; Mannherz, H. G.; Suck, D.; Pai, E. F.; Holmes, K. C. *Nature* **1990**, *347*, 37.
299. Zahm, J. A.; Padrick, S. B.; Chen, Z.; Pak, C. W.; Yunus, A. A.; Henry, L.; Tomchick, D. R.; Rosen, M. K. *Cell* **2013**, *155*, 423.
300. Irobi, E.; Aguda, A. H.; Larsson, M.; Guerin, C.; Yin, H. L.; Burtnick, L. D.; Blanchoin, L.; Robinson, R. C. *EMBO J.* **2004**, *23*, 3599.
301. Bertazzon, A.; Tian, G. H.; Lamblin, A.; Tsong, T. Y. *Biochemistry* **1990**, *29*, 291.
302. Contaxis, C. C.; Bigelow, C. C.; Zarkadas, C. G. *Can. J. Biochem.* **1977**, *55*, 325.
303. Le Bihan, T.; Gicquaud, C. *Biochem. Biophys. Res. Commun.* **1993**, *194*, 1065.
304. Turoverov, K. K.; Biktashev, A. G.; Khaitlina, S. Y.; Kuznetsova, I. M. *Biochemistry* **1999**, *38*, 6261.
305. Tatunashvili, L. V.; Privalov, P. L. *Biofizika* **1984**, *29*, 583.
306. West, J. J.; Nagy, B.; Gergely, J. *The Journal of biological chemistry* **1967**, *242*, 1140.
307. Kuznetsova, I. M.; Khaitlina, S.; Konditerov, S. N.; Surin, A. M.; Turoverov, K. K. *Biophys. Chem.* **1988**, *32*, 73.
308. Turoverov, K. K.; Kuznetsova, I. M.; Khaitlina, S. Y.; Uverskii, V. N. *Prot. Pept. Lett.* **1999**, *6*, 73.
309. Turoverov, K. K.; Verkhusha, V. V.; Shavlovsky, M. M.; Biktashev, A. G.; Povarova, O. I.; Kuznetsova, I. M. *Biochemistry* **2002**, *41*, 1014.
310. Kuznetsova, I. M.; Stepanenko, O. V.; Stepanenko, O. V.; Povarova, O. I.; Biktashev, A. G.; Verkhusha, V. V.; Shavlovsky, M. M.; Turoverov, K. K. *Biochemistry* **2002**, *41*, 13127.
311. Povarova, O. I.; Kuznetsova, I. M.; Turoverov, K. K. *Tsitologiia* **2005**, *47*, 953.
312. Altschuler, G. M.; Klug, D. R.; Willison, K. R. *J. Mol. Biol.* **2005**, *353*, 385.
313. Turoverov, K. K.; Kuznetsova, I. M.; Uversky, V. N. *Prog. Biophys. Mol. Biol.* **2010**, *102*, 73.
314. Kuhn, T. S. *The Structure of Scientific Revolutions*; Fourth ed.; University Of Chicago Press: Chicago, 2012.
315. van der Lee R., Buljan M., Lang B., Weatheritt R.J., Daughdrill G.W., Dunker A.K., Fuxreiter M., Gough J., Gsponer J., Jones D.T., Kim P.M., Kriwacki R.W., Oldfield C.J., Pappu R., Tompa P., Uversky V.N., Wright P.E., Babu M.M. (2014) Classification of intrinsically disordered proteins and regions. Chemical Reviews. 114 (7) In press